BASIC BIOLOGY COURSE
UNIT 2
ORGANISMS AND THEIR ENVIRONMENT

BOOK 4

Ecological Principles

MICHAEL A. TRIBE, MICHAEL R. ERAUT &
ROGER K. SNOOK
University of Sussex

CAMBRIDGE UNIVERSITY PRESS

CAMBRIDGE
LONDON · NEW YORK · MELBOURNE

Published by the Syndics of the Cambridge University Press
The Pitt Building, Trumpington Street, Cambridge CB2 1RP
Bentley House, 200 Euston Road, London NW1 2DB
32 East 57th Street, New York, NY 10022, USA
296 Beaconsfield Parade, Middle Park, Melbourne 3206, Australia

©Cambridge University Press 1975

Library of Congress catalogue card number: 75–6285

ISBNs:
    0 521 20658 8 hard covers
    0 521 20638 3 limp covers

First published 1975

Printed in Great Britain
at the University Printing House, Cambridge
(Euan Phillips, University Printer)

BOOK 4

Ecological Principles

Basic Biology Course

Unit 1   Microscopy and its Application to Biology
Book 1 *Light Microscopy*
Book 2 *Electron Microscopy and Cell Structure*
Book 3 *Dynamic Aspects of Cells*
    Film strip for Book 2
    Tape commentary (cassette) for Book 2
    4 super-8 film loops for Book 3

Unit 2   Organisms and their Environment
Book 4 *Ecological Principles*
The Ecology Game pack
    Film strip for Book 4
    Film strip for Ecology Game

Unit 3   Regulation within Cells
Book 5 *Cell Membranes*
Book 6 *Photosynthesis*
Book 7 *Enzymes*
Book 8 *Metabolism and Mitochondria*
Book 9 *Protein Synthesis*
The Enzyme Game pack
    Super-8 film loop for Book 5
    Film strip for Book 6 and Book 8
    Tape commentary (cassette) for Book 6 and Book 8

Unit 4   Communication between Cells
Book 10 *Nerves*
Book 11 *Hormones*
    Film strip for Book 10
    Film strip for Book 11

Tutors' Guide

Contents

Foreword

This book is part of a Basic Biology Course for undergraduates written by the Inter University Biology Teaching Project Team at Sussex.

The main aim of the book is to provide you with an understanding of the structure of ecosystems. An ecosystem may be defined as a unit within the area in which life is possible on our planet, where living and non-living things interact, and where, with the aid of energy from sunlight, both living and non-living materials are continuously recycled — i.e. a 'self-contained' unit. We feel strongly that all people who call themselves 'educated' should know about the biological nature of the world in which they live. We hope that you will share our belief at the end of this book, because at no time in the history of man has it been so important for all of us to possess this understanding. Unless this and future generations can appreciate what actual and potential havoc the increasing population of man can create, and in turn take measures to avert it, we may well be heading on the road, not only to self-destruction, but also the destruction of other living organisms which are part of our environment.

You will see that the emphasis of the book is on the dynamic structure of ecosystems. There is virtually no mention of the genetic and evolutionary principles behind such concepts as selection pressure, adaptation and speciation. This has been quite intentional, as these concepts will form part of the subject matter of a separate course (the Joint Universities Genetics Course) in which the University of Sussex is working collaboratively with the Open University and the Universities of Birmingham and Hull. It is intended that this genetics course will be available to students in 1976.

Brighton, Sussex, 1974
M.A. Tribe
M.R. Eraut
R.K. Snook

Acknowledgements

This book was developed under the auspices of the Inter University Biology Teaching Project and is the responsibility of the Sussex University Project Team. However, it owes a great deal to the students who studied and criticized our earlier versions and to many colleagues both at Sussex and elsewhere who made constructive suggestions for its improvement.

In particular we would like to thank the following:

The Nuffield Foundation for financially supporting the project from 1969-72;

Cambridge University Press for the continued interest and support in publishing the materials;

Mrs P. Smith and Mrs S. Collier, the project secretaries;

Mr Colin Atherton for photographic assistance; and

Mr D. Streeter (University of Sussex) and Professor W.H. Dowdeswell (University of Bath) for their advice.

4.0. Introduction

One of the effects of our highly technological, functionally specific society, has been to make us 'consumer minded'. That is to say, we buy a product such as meat or bread or a car, with little thought about who produces it, how it is produced, or what the side effects of its production are.

Unfortunately, the manufacture of cars involves the deposition of vast quantities of slag on otherwise useful agricultural land, the evolution of poisonous vapours into the atmosphere, and extensive urbanization of the countryside. With the indiscriminate spread of industries such as these, less and less land is available for the rearing of cattle and the cultivation of wheat. Intensive agricultural methods are therefore employed, involving the massive use of inorganic fertilizers, insecticides, deforestation, and the production of poor-quality foods. These man-made assaults on nature, if uncontrolled, will take from the industrial society, both its source of food, and the amenity of an unspoiled environment. It is only by understanding the patterns and balances of nature that we can intelligently organize our production activities, control our consumer demands, and avoid the prospect of being surrounded by a sterile, featureless wasteland.

Another effect of 'consumer mindedness' has been to insulate us from the 'Third World', where starvation and disease are the order of the day. Often such conditions obtain because of ignorance, and the consequent inability to co-operate with nature to the advantage of man. It therefore seems to be incumbent on the more fortunate societies to help the starving majority by giving them an understanding of the patterns and balances of nature, and by providing them with the technology to maximize their natural resources, reclaim marginal land, deserts and marshes, and, one day perhaps, farm the sea.

4.1. Interactions between organisms and their non-living environment

4.1.1. Discussion

One of the effects of pollution on the environment has been to change it to such a degree, and at such a rate, and in such a direction, that many living things are unable to adjust to the change. It follows that if we are to manage our impact on the environment, we must first of all have an understanding of the requirements which living things have of their surroundings.

We shall start therefore by looking at a geographically well defined area — a lake — and see how living things in the lake distribute themselves. We shall then try to discover what variables in the lake are the determining factors in this distribution.

4.1.2. Overview and objectives

In this section we shall look at the evidence for the dependence of cells and organisms on various inorganic substances and energy sources in the environment. The products of this interaction in the form of new living material will be considered.

At the end of this section you should be able to:
1. Formulate and explain the concept of limiting factors.
2. Formulate and explain the concept of production.
3. Demonstrate some skill in
   (*a*) formulating hypotheses;
   (*b*) designing experiments;
   (*c*) interpreting graphical data.
4. List and describe the basic requirements which plants have of their environment.

4.1.3. Preknowledge requirements

(i)   The general structure of plant and animal cells as revealed by the light microscope, e.g. cell wall, nucleus, chloroplast, mitochondria, vacuole.
(ii)   The chemical symbols for the following chemical elements of the periodic table, e.g. C, H, O, N, S, P, Fe, Mn, Mg, Mo, S.
(iii) SI units: metres (m), grammes (g), litres (l), seconds (s).
(iv)   Elementary understanding of chemical isotopes.
(v)   Some knowledge of the chemical properties of O_2, CO_2 and H_2O; oxidation and reduction.

4.1.4. Instructions on working through the programmed sections

In the programmed sections which follow, questions and answers are arranged sequentially down the page. You are provided with a masking card and a student response booklet. Cover each page of the book in turn (unless otherwise instructed) and move the masking card down to reveal two thin lines:

This marks the end of the first question on that page. Record your answer to the question under the appropriate section heading in the response booklet provided. Then *check* your answer with the answer given. If your answer is correct, move the masking card down the page to the next two thin lines and so on. If any of your answers are incorrect retrace your steps and try to find out why you answered incorrectly. If you are still unable to understand the point of a given question, make a note of it and consult your tutor. The single thick line

---

is a demarcation between one frame and the next.
Bold double lines signify convenient stopping points in the book.

4.1.5. Resource materials required

Before starting the section make sure you have the following materials:

beaker of pond water	iodine solution
1 microscope	petri dish
microscope slides	bunsen burner
coverslips	test tube holder
a needle	coarse forceps
test tubes	access to a variegated (green and white leaved) plant
dropping pipettes	access to a slide projector and set of slides

Examples of algae

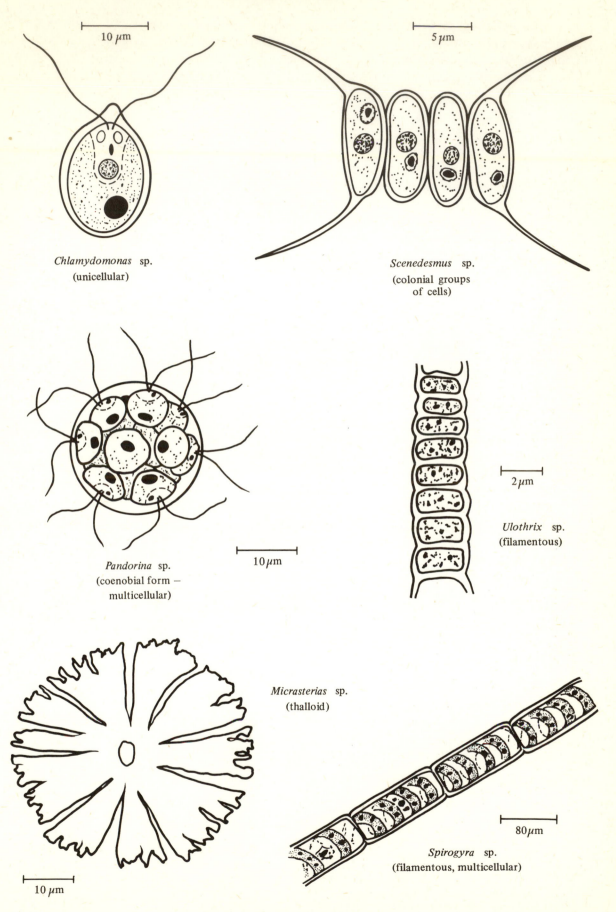

10 μm

Chlamydomonas sp.
(unicellular)

5 μm

Scenedesmus sp.
(colonial groups
of cells)

Pandorina sp.
(coenobial form —
multicellular)

10 μm

2 μm

Ulothrix sp.
(filamentous)

Micrasterias sp.
(thalloid)

80 μm

Spirogyra sp.
(filamentous, multicellular)

10 μm

4.1.6. Looking into lakes and ponds

1   In the beaker marked A on the bench in front of you is a sample of water containing organisms taken from a pond. Using the dropping pipette provided, place a drop of this water in the depression of the cavity slide provided. Examine the drop under the microscope, using low- and high-power objectives. Give two reasons why you believe the objects you can see are living.

(*a*)   They have a cellular organization.
(*b*)   They show autonomous movement (some).

2   Can you suggest other criteria you would use to confirm your conclusion?

(*a*)   Growth
(*b*)   Reproduction

3   Are the organisms you can see uni- or multi-cellular?

Some are single-celled — uni-cellular (or acellular).
Some are long chains of cells — filamentous. ⎤
Some are sheets of cells — thalloid. ⎦ Multicellular

(See diagrams opposite.)

4   What colour are the individual cells of these organisms?

Green or greenish yellow

5   Do the cells of these organisms have cell walls?

Yes

6   Would you say that these organisms are plants or animals? Give your reasons.

_____

_____

Plants, since they are green and have cell walls.

7   How many different types of unicellular plants can you see?

_____

_____

There may be many different types.

8   These minute plants are called *phytoplankton* (*phyto* = plant; *plankton* = floating or drifting life) and are found distributed in the main body of the lake water. If you wanted to find out which parts of a lake are most favourable to the phytoplankton population in early April — a time of rapid growth — what measurements would you make?

_____

_____

You probably suggested measuring the distribution of phytoplankton in samples of lake water; either samples from different points on the lake's surface, or samples from different depths, or both. In fact, depth is the critical factor for reasons which will emerge later.

9   Can you suggest why the distribution of phytoplankton should not be measured at a single point in time?

_____

_____

It would be impossible to distinguish the effects of past conditions in the lake from those of the present.

10   Bearing in mind your answer to frame 2, which of the following distributions would you use to ascertain the most favourable parts of the lake for phytoplankton? (Give your reasons.)
(a) Distribution of the increase in weight of living phytoplankton over a given period.
(b) Distribution of living phytoplankton.
(c) Distribution of the increase in the number of living phytoplankton over a given period.

_____

_____

    a, which takes both growth and reproduction into consideration, would give the most accurate information. This quantity being a measure of the total increase in living material over a given period of time is given the name *production*.

    b may be the result of conditions which no longer obtain.

    c does not take growth into consideration.

---

11 All cells contain highly variable but relatively large quantities of water. Most of this water acts as a chemically inert medium supporting the chemical processes of life. So in what form would you weigh the phytoplankton?

Dehydrated. In this way one would measure the increase in dry weight, i.e. only the new *living* material. This quantity is called the *dry weight production*, though we often refer to it simply as the production.

---

12 The distribution of phytoplankton dry weight production with respect to depth has been measured and is inserted on the graph below. Where would you say that conditions were most favourable for phytoplankton?

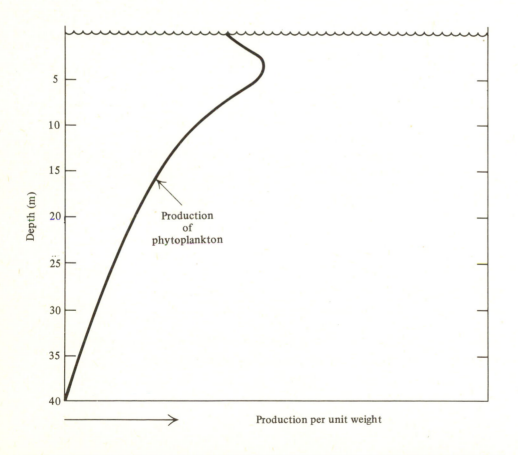

Production per unit weight

At a depth of about five metres

13 Why has production *per unit weight* been chosen for the horizontal axis?

Because it eliminates the concentration of phytoplankton as a variable; and this might depend on past as well as present conditions in the lake.

14 If W_1 and W_2 are the dry weights of phytoplankton in samples of equal volume taken at the same depth at the beginning and end of the experimental period, what is the production per unit weight at that depth?

$$\frac{W_2 - W_1}{W_1}$$

15 A chemical analysis of living things reveals that the elements generally found in greatest quantity are C, H, O, N, S and P. Other elements are present to a greater or lesser extent. Clearly, however, living things are more than a heap of chemicals. In fact, the characteristic which distinguishes living organisms from other non-living chemical systems is their supremely high level of complexity and sophisticated organization. Yet, as we have seen, in order to maintain themselves, living organisms depend on the presence of a favourable environment.
In the light of this discussion, what general environmental parameters, both physical and chemical, may have determined the distribution of production shown above?

Physical factors might be temperature, light, pressure.
Chemical factors might be the concentrations of various molecules containing C, H, O, N, S and P.

16 If several of the above factors influence production, which of them is likely to determine the shape of the production curve?

The one which is in too short or too great a supply. This is called the *limiting factor*.

17 Could there be more than one limiting factor responsible for our production distribution curve?

Yes

18 Let us first consider the possible effect on phytoplankton distribution of variations in the concentration of inorganic salts of nitrogen, sulphur and phosphorus. Examine the graph below.

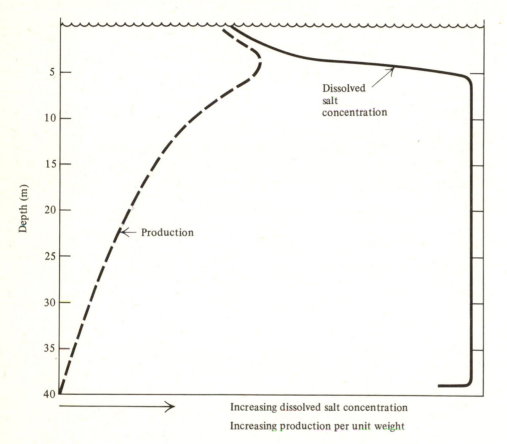

In what ways might this inorganic salt distribution be responsible for aspects of the production curve (dashed line)? Suggest two hypotheses.

(a) Low salt concentration may reduce production at depths of less than five metres.
(b) High salt concentration may reduce production at depths of more than five metres.

19 **Slide 1** shows three beakers, one dark green (high production) containing x g/l of salts; the second medium green (medium production) containing ½x g/l of salts; and the third light green (low production) contains ¼x g/l of salts. All three concentrations were higher than those found anywhere on the lake. The green colour results from the presence of phytoplankton grown for four weeks in the solutions. What information does the experiment give you with respect to the above hypotheses?

_____

_____

(i) Since production (as indicated by colour intensity) decreases with the decrease in salt concentration, the experiment gives further support to hypothesis *a*.

(ii) The high production levels observed in salt concentrations higher than those found in the lake make hypothesis *b* extremely unlikely.

20 To confirm these results in a real environment requires a field experiment. One such experiment, which attempts to separate the effects of physical factors from those of chemical factors, is described below.

Samples of lake water were taken from various depths in bottles, two from each depth. The dry weight of phytoplankton at the beginning was determined from the first of each pair of bottles, while the second was suspended just below the surface for a period of time before its dry weight was also determined. The production per unit weight, i.e.

$$\frac{W_2 - W_1}{W_1}$$

was then calculated from each pair of samples and plotted against the depth from which the samples originally came.

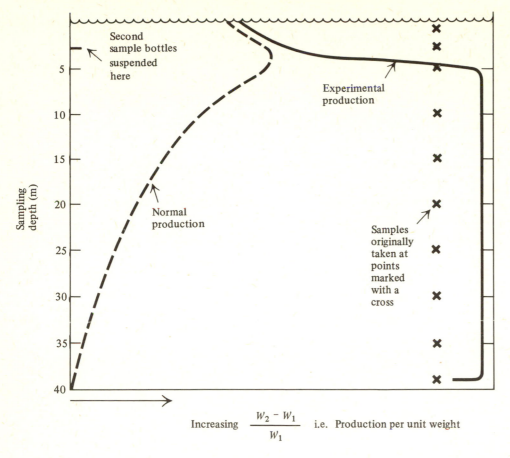

For each sample bottle suspended just below the surface the factors corresponding to surface conditions were_____, and the factors corresponding to conditions at the sample depths were

_____.

(Choose from temperature, light, pressure, and chemical factors.)

Surface conditions: *temperature, light*
Sample depth conditions: *pressure, chemical factors*

21 How do the results of this experiment affect the two hypotheses formulated in frame 18? Explain your answer.

There is some support for hypothesis *a*, as the drop in production in going from 5 m samples to surface samples can only be caused by chemical factors. (However, there could still be factors other than salt concentration, so hypothesis *a* is not definitely proved.) Hypothesis *b* is conclusively disproved as the chemical conditions in the samples from below 5 m clearly do not limit production.

22 It thus appears that physical factors must be responsible for the fall in production below the 5 m level under normal conditions. Can you suggest an experiment which might give further support to this conjecture?

(*Hint.* It uses the opposite procedure to the last experiment.)

Samples of lake water are taken from just below the surface and then suspended at different depths for a period of time. (Since all the bottles are originally from the same sample depth, only one is needed to establish the initial dry weight of phytoplankton.)

23 The results of this experiment are shown below. What can you conclude?

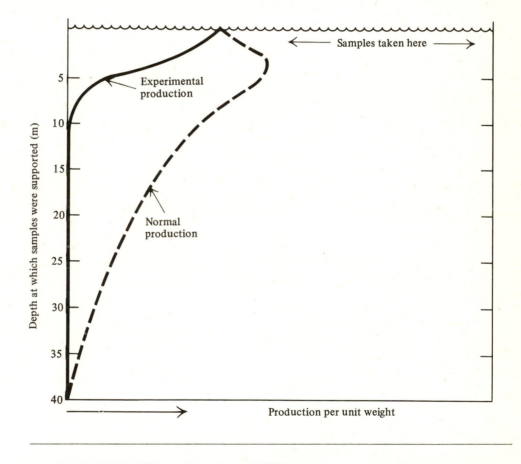

That physical factors alone can account for the fall in production below the 5 m level under normal conditions.

The normal production curve (dashed line) can be explained by the combination of physical factors (solid line in frame 23) and chemical factors (solid line in frame 20).

24 Let us now examine the distribution curves of temperature and light. Do these suggest that either or both might be limiting below 5 m?

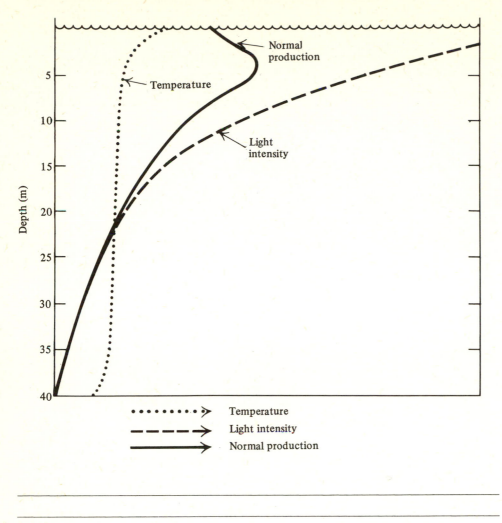

Yes. Low temperature and low light may both be limiting below 5 m.

25 To distinguish between the effect of temperature and the effect of light, the experiment described in frame 20 was repeated in modified form. Four bottles were taken from each sampling depth and treated as follows:

Set 1.   Used to determine the initial dry weight of phytoplankton.

Set 2   Enclosed in a transparent thermostatically controlled
(cold   container set to maintain the temperature as cold as that at
set).   40 m; and then suspended just below the surface.

Set 3   Darkened to give the light intensity normally found at 40 m;
(dark   and then suspended just below the surface.
set).

Set 4   Suspended just below the surface, as in frame 20.
(control
set).

The results are shown overpage. What can you conclude about the effects of light and temperature?

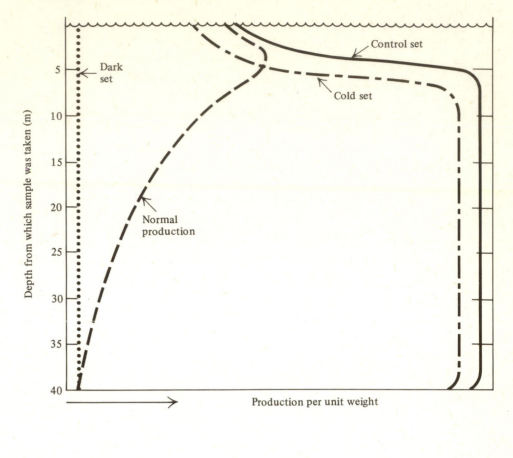

Light is the main limiting factor below 5 m.
Temperature is only limiting to a small extent.

26   Although we can now explain the general shape of the normal
      production distribution in terms of light and salt concentration, we
      have not eliminated the possibility of other chemical factors being
      limiting. We know that growth requires C, H and O, and water is the
      only source of these elements which we know will not be limiting.
      Both CO_2 and O_2 are obvious candidates for investigation.
      Consider the following laboratory experiment, designed to investigate
      the fate of the oxygen atoms. The two most plentiful sources of oxygen
      are water (H_2O) and molecular oxygen (O_2). In order to distinguish
      between the two sources, the rare isotope of oxygen ^{18}O was used, so
      that some of the molecular oxygen dissolved in the water had a mass of
      34 (one atom of ^{16}O and one atom of ^{18}O) but the water itself only
      contained the normal ^{16}O.
      If new molecular oxygen was being produced from water by living
      organisms its mass would be_____, but if molecular oxygen was
      being consumed by living organisms, its mass would be either_____
      or_____.

32; because there is no ^{18}O oxygen combined as H_2O.
32 or 34; because the molecular oxygen is a mixture of masses 32 and 34.

27 In the experiment a sample of phytoplankton from the 5 m depth in the lake was saturated with molecular oxygen of mass 34; and then suspended under conditions similar to those normally met at 5 m depth in the lake. It was then treated to various light conditions while the changes in concentration of molecular oxygen of both mass 32 and mass 34 were observed. The results are shown below.

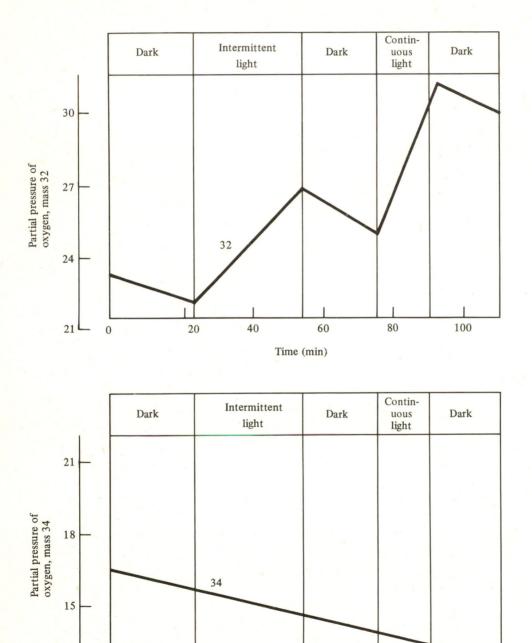

15

    (a)   Does the production rate of molecular oxygen depend on the light conditions?

    (b)   Does the consumption rate of molecular oxygen depend on the light conditions?

    (a)   Yes. It does not occur at all in the dark.

    (b)   No, because the mass 34 graph is a straight line.

---

28   Is oxygen of mass 32 being consumed as well as produced during the light periods?

Yes, at much the same rate as the oxygen of mass 34. (The phytoplankton will not know which is which!)

---

29   In the light of the information in frames 26 to 28, summarize (in two sentences) what you know about the uptake and release of oxygen by phytoplankton.

    (a)   Irrespective of the presence or absence of light, molecular oxygen dissolved in the water is absorbed by phytoplankton.

    (b)   In the presence of light, phytoplankton convert oxygen atoms combined in water into molecular oxygen.

(Since this latter process must involve the 'splitting' of hydrogen-oxygen bonds in water molecules, it is called *photolysis* which literally means 'light-splitting'.)

Note. We look at photolysis in more detail in Book 6 (Unit 3).

---

30   Under what conditions in the lake might one of these processes become dependent on the other?

Under conditions of very low oxygen supply, process *a* would depend on process *b* for its oxygen supply.

---

31   Assuming no other source of molecular oxygen than process *b*, under what circumstances would the concentration of dissolved oxygen remain constant?

When the rate at which oxygen is released by process *a*, is the same as that at which it is taken up by process *b*. Such a state is described as the *compensation point* of the two processes.

32　Transparent and blackened bottles were suspended at different depths in the lake. Each bottle contained water taken from the 5 m depth and the same quantities of the phytoplankton found at that depth. At the start of the experiment, the oxygen concentration in each bottle was measured, and again at the end of the experiment. The results are shown below.

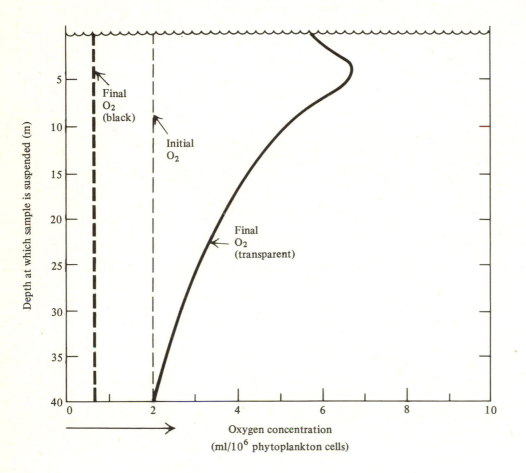

(*a*)　What processes are affecting the oxygen concentration in the blackened bottles?

(*b*)　What processes are affecting the oxygen concentration in the transparent bottles?

(*c*)　At what depth is the compensation point?

(*a*)　Only oxygen uptake

(*b*)　Both oxygen uptake and oxygen release

(*c*)　At 40 m, because there is no change in concentration

---

33 Approximately how much oxygen was evolved at this depth during the experiment?

_____

_____

1.25 ml/10^6 phytoplankton cells, because that was the amount absorbed in the darkened bottles.

34 We have established that photolysis of water occurs with release of oxygen as O_2. The question now arises as to what happens to the hydrogen that remains. The graph below shows the result of an experiment in which normal water was mixed with increasing percentages of deuterium or heavy water (D_2O). Phytoplankton were grown in the different mixtures for four days and their production at the end of that period was measured. What does the graph tell us about the importance of hydrogen from water for plant production?

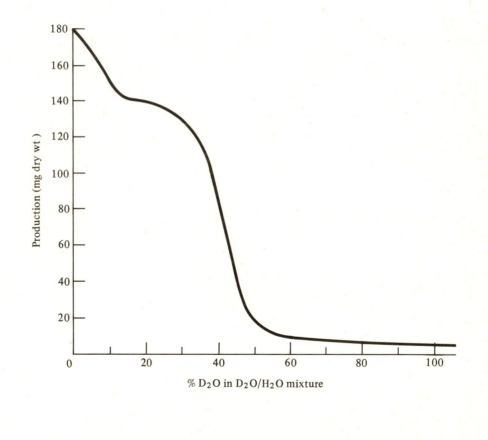

% D_2O in D_2O/H_2O mixture

_____

_____

It is essential and it has to be H rather than D.

35 Let us now look at a mass spectrogram which traces the changes in dissolved CO_2 concentration in the frame 27 experiment. How do the phytoplankton change the concentration of dissolved CO_2 in the dark?

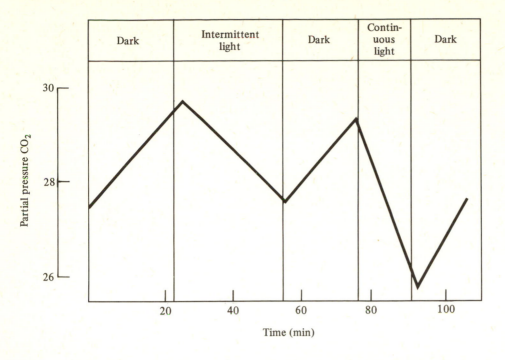

They give out CO_2, i.e. they increase the concentration.

36 And in the light?

They take up CO_2, i.e. they decrease the concentration.

37 Which of the following hypotheses still need to be tested?
  (*a*) Carbon dioxide is released in the light.
  (*b*) Carbon dioxide is taken up in the light.
  (*c*) Carbon dioxide is released in the dark.
  (*d*) Carbon dioxide is taken up in the dark.

a and *d*

38 Design an experiment using the ^{14}C isotope of carbon to test hypothesis *d*. Do not attempt to go into too much detail — two or three sentences will suffice.

Place two identical samples in media containing dissolved $^{14}CO_2$. Place one in the dark and the other in the light (control). Record $^{14}CO_2$ uptake in each (if any).

---

39 The graph below shows the uptake of $^{14}CO_2$ in the dark and in the light. Does this support the hypothesis?

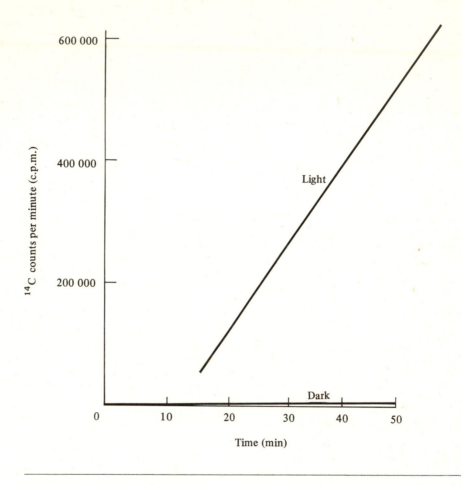

Yes; though the uptake in the dark is negligible when compared with that in the light, and can be ignored as a significant mechanism for carbon intake.

---

40 Design a similar experiment using the ^{14}C isotope of carbon to test the final remaining hypothesis: that carbon dioxide is released in the light.

Allow two identical samples of phytoplankton to take up $^{14}CO_2$ in the light, then blow unused $^{14}CO_2$ out of the samples with $^{12}CO_2$. Place one sample in the light for several hours, and the other in the dark (control). Record the $^{14}CO_2$ release every two hours (if any).

---

41 The results of this experiment are shown below. Is CO_2 released in the light?

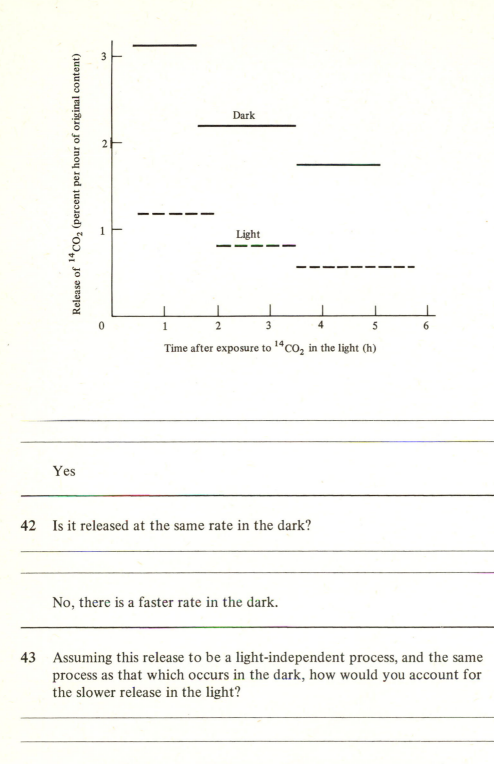

Time after exposure to $^{14}CO_2$ in the light (h)

Yes

42 Is it released at the same rate in the dark?

No, there is a faster rate in the dark.

43 Assuming this release to be a light-independent process, and the same process as that which occurs in the dark, how would you account for the slower release in the light?

The lower rate of CO_2 release in the light could be accounted for by the 'CO$_2$-absorbing process' which trapped the CO_2 before it left the immediate environment of the phytoplankton cells. In other words there would be a higher probability of absorbing a $^{14}CO_2$ molecule from the surrounding solution. This is essentially what happens.

44 We have now established the following exchanges between the phyto-plankton and their environment.

(*a*)   Certain salts, in particular, salts of N, S, and P are required.

(*b*)   *Light-dependent interaction.* In the presence of light, photolysis of water occurs, the oxygen being released to the atmosphere and the hydrogen used by the plant. At the same time CO_2 is taken up. This interaction is called *photosynthesis.* (It will be discussed in more detail in Book 6, Unit 3.)

(*c*)   *Light-independent interaction.* Irrespective of the presence or absence of light, dissolved molecular oxygen is taken up, and CO_2 released. This interaction is called *respiration.* (It will be discussed in more detail in Book 8, Unit 3.)

The question now arises as to whether oxygen and carbon dioxide are ever in limiting supply in the lake itself. Examine the graphs below which show the distribution of CO_2 and O_2 in relation to production.

Bearing in mind what you already know, and assuming no other organisms to be involved, how would you account for:

(*a*)   The low CO_2 and high O_2 at 5 m depth?

(*b*)   The high CO_2 and low O_2 at 40 m depth?

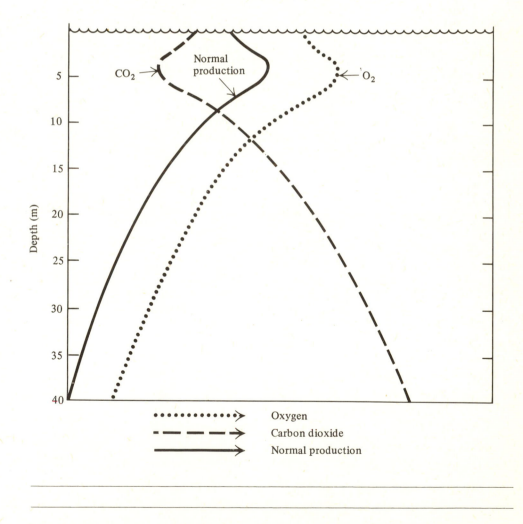

a is due to the predominance of photosynthesis over respiration at 5 m.
b is due to the predominance of respiration over photosynthesis at 40 m.

45 Assuming that CO_2 is not present in *too great* a quantity at any depth, at what depth is it most likely to be limiting on production?

Around 5 m, since at that depth CO_2 is being drastically depleted by photosynthesis.

46 A modified form of the 'suspended bottle' experiment can be used to test this hypothesis. Several bottles are taken from each sampling depth and then one set is used for determining the initial dry weight of phytoplankton at each depth. The other sets are enriched with CO_2 and then suspended at the sampling depths of their contents, each set having a given initial concentration of dissolved CO_2. Model results of such an experiment are given below. The numbers represent the level of CO_2 concentration. The solid lines show normal production, the dashed lines show light intensity and the dotted lines show production with added CO_2. Is CO_2 in limiting supply at around 5 m depth?

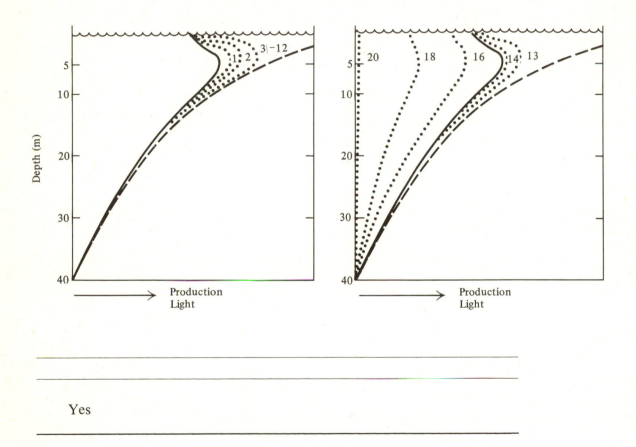

Yes

47 What is the optimum concentration range of carbon dioxide?

3–12

48 What is limiting production between concentrations 3–12 of carbon dioxide?

Light intensity below 5 m and salt concentration above 5 m.

49 What do you think from your knowledge of the properties of CO_2 in solution may be limiting production between concentrations 13 to 20 of carbon dioxide?

Decreasing pH (increasing acidity)

50 This is in fact the case. Most pH distribution curves in the lake are the same shape as CO_2 curves. What explanation does this suggest for the slight depression of production at 40 m in the experimental results of frame 20?

It could be due to the low pH.
(This is the reason but the low pH is only partly due to high CO_2 concentration.)

51 Since our primary concern is production, the question that now arises is this. In what form are the CO_2 and hydrogen incorporated as living material in the plant? To answer this question, we again resort to the use of ^{14}C. Phytoplankton cells are placed in a medium containing $^{14}CO_2$, exposed to light for five seconds and immediately killed. Then their contents are analysed by two-dimensional paper chromatography. The principles of this process are that (i) different substances have different solubilities in various solvents, and (ii) substances differ in the extent to which they are adsorbed onto solid surfaces such as paper. In paper chromatography, a very small quantity of a mixture is deposited as a spot at one corner of a piece of absorbent paper. This spot is referred to as the origin. The paper is then dipped into a carefully selected solvent along one edge adjacent to the origin. Just before the solvent front has reached the far end of the paper, the paper is removed and dried. The paper is turned through $90°$ and the process repeated in a second solvent. The final position of any one component in the original mixture is determined by its solubilities in each of the solvents and its relative adsorbance to the paper. When radioactive isotopes are used, the positions of colourless chemical compounds can be detected by leaving the paper chromatogram on a photographic film for an appropriate period. The developed film is shown below, and is called an autoradiograph or radiogram. What do the black spots represent?

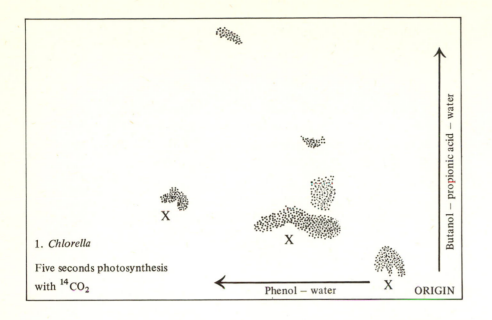

1. *Chlorella*

Five seconds photosynthesis
with $^{14}CO_2$

Butanol – propionic acid – water

Phenol – water

ORIGIN

Substances into which the radioactive carbon dioxide was incorporated

52   Subsequent analysis of these substances revealed that they were sugar phosphates and were readily hydrolysed to give sugars and phosphate ions. Sugars have the general formula $(CH_2O)_n$ (where n is 5, 6 or 7) and therefore belong to that class of compounds which we call carbohydrates (carbo, C, + hydrate, H_2O = carbohydrate, CH_2O). The formation of these sugars involves the reduction of CO_2 by hydrogen. In the typical case of glucose such a reduction has been shown to require an energy input of 2880 kilojoules (kJ). In view of the evidence presented previously, where do you think this energy comes from?

Light

53 The next autoradiograph was made using phytoplankton which had
been left in $^{14}CO_2$ for 4 min. The sugars are again marked. A new class
of substances has now appeared called amino acids, so-called because
they have the general formula

$$NH_2 - \underset{R}{CH} - COOH$$

where R represents one of twenty possible carbon chains. The seven of
the twenty visible on our chromatogram are numbered 1 to 7. As a
result of comparing chromatograms 1 and 2, what is a likely immediate
source of the radioactive carbon in the amino acids?

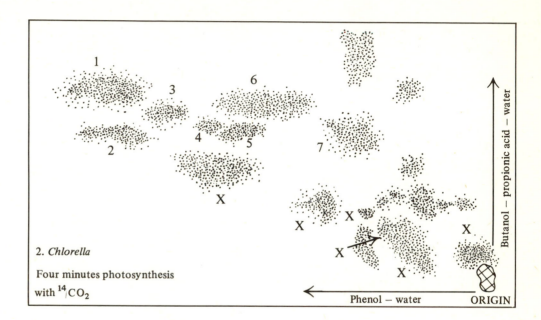

2. *Chlorella*

Four minutes photosynthesis
with $^{14}CO_2$

Phenol – water   ORIGIN

Butanol – propionic acid – water

_____

_____

Sugars

_____

54 How would you confirm this?

_____

_____

Feed radioactive sugars to the plant and see if the label enters newly
formed amino acids

_____

55 Sugar phosphates are the compounds first formed in the light; and they can be converted to amino acids or polymerized to form polysaccharides. The amino acids can also be polymerized to form proteins. Redraw the diagram below and fill in the arrows tracing the steps from raw materials to end products such as proteins and polysaccharides. Five more arrows are needed.

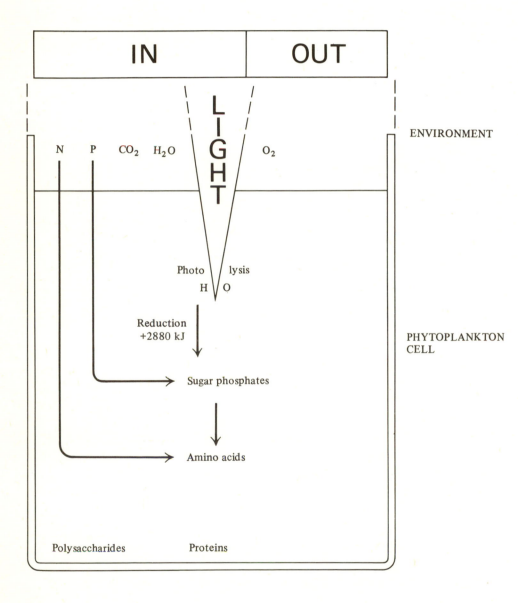

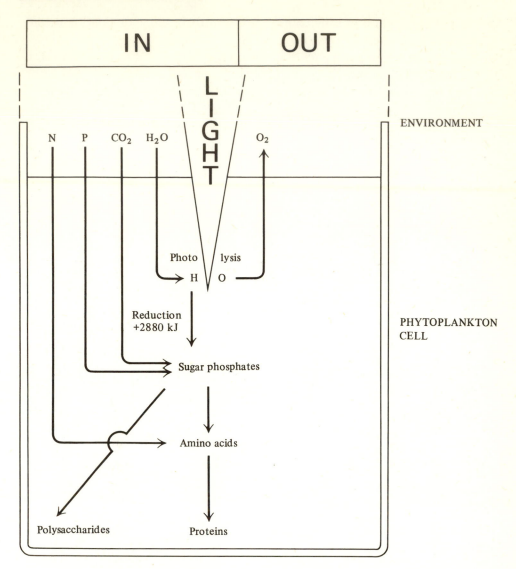

Note the incorporation of P and N into sugars and proteins respectively. These are by no means the only substances containing N and P, but the inclusion of this information on the flow diagram illustrates the essential nature of these two elements.

---

56 We have established that sugars are formed as a result of the light-dependent fixation of CO_2. This process is called photosynthesis, and in each sugar molecule formed by photosynthesis 2880 kJ of energy are trapped.
How could this energy be made available to the cell?

By reconversion to CO_2 and H_2O, i.e. reversing the photosynthetic process

---

57 What would the products of this process be?

CO$_2$, water and energy (2880 kJ)

58 If one assumes dissolved oxygen to be the ultimate oxidizing agent, do our discoveries about the light-independent process as summarized in frame **44c** correspond to the above predictions?

Yes

59 Energy is therefore made available for the multifarious life processes of the cell. Using double lines, insert on your answer to frame 55 the information you have acquired about the light-independent process called respiration. (You will need to insert O$_2$ on the *in*side, four sets of double lines, and 2880 kJ as a product.)

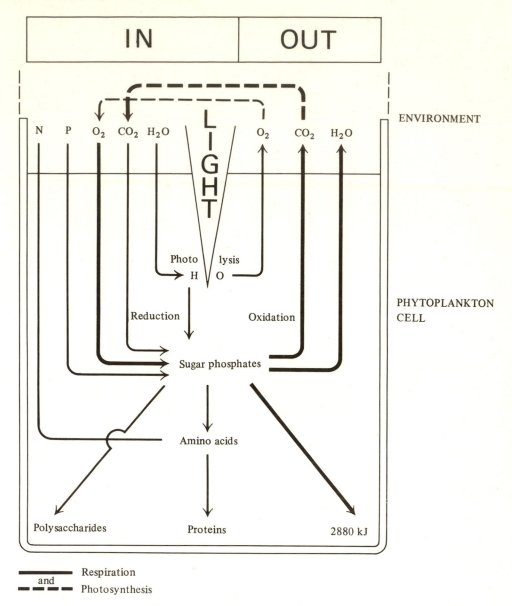

In fact, just under half of the 2880 kJ is retained by the organism, and the rest is lost to the environment.

---

60 Starch, a polysaccharide, is a polymer of the sugar glucose, which is the sugar formed in photosynthesis. This sugar is also used to supply energy in respiration. **Slides 2 and 3** show identical filamentous plankton cells stained with iodine to reveal their starch content. They illustrate the response of this particular planktonic organism to the following conditions:

   Slide 2.   At the beginning of a light period after a dark period or at the end of a dark period.
   Slide 3.   At the end of a light period.

   Note. Blackish colouration indicates the presence of starch.

   What does a comparison of these two slides reveal about the function of starch with respect to photosynthesis and respiration?

Excess sugar made in the light is polymerized to form starch, which is then depolymerized in the dark to supply the sugar needed in respiration.

---

61 We can now turn to the question, how does sunlight succeed in splitting water in the phytoplankton cell? Let us conduct an experiment using variegated (white and green) leaves. Take a green and white leaf from the plant on the bench. Place it in a test tube of boiling water and boil for 2 min. (*Caution* — do not damage the leaf when inserting or removing from test tube.) Remove the leaf and lay it flat in a petri dish of iodine. Make sure that it is completely submerged. Leave for 5 min. What do the results of this experiment suggest might be the active element in trapping light for the photolysis of water?

The pigment — in this case the green pigment chlorophyll

---

62 Insert chlorophyll on the flow diagram of frame 59.

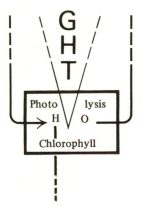

---

63 As we have seen in frames 53 and 55, nitrogen is essential in the formation of amino acids, which in turn form proteins — the main structural molecules of the cell. If nitrogen is fed to phytoplankton in the form of nitrates, clearly the nitrates must be reduced to form amino acids

$$(\text{formula } NH_2 - \underset{R}{CH} - COOH)$$

Now consider the following graphs. These show the effect of *molybdenum* on production of *Scenedesmus* (a single-celled phytoplankton organism) supplied with nitrates and ammonia. What conclusions do you draw as to the importance of molybdenum to the plant cell?

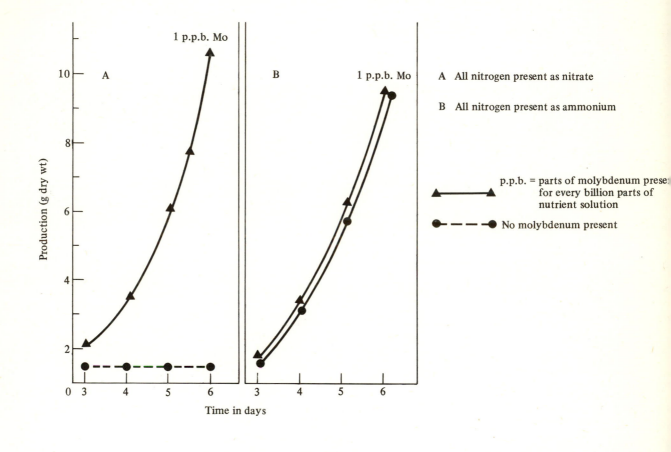

A All nitrogen present as nitrate

B All nitrogen present as ammonium

p.p.b. = parts of molybdenum prese[nt] for every billion parts of nutrient solution

No molybdenum present

_____

_____

It is essential to the reduction of nitrate to ammonium ions and therefore amino acid and protein production.

_____

64 Consider another element which occurs in minute quantities in lakes — namely manganese. If manganese is essential for photosynthesis, what will be the effect on graph B below of adding manganese after 40 min to the phytoplankton culture lacking manganese?

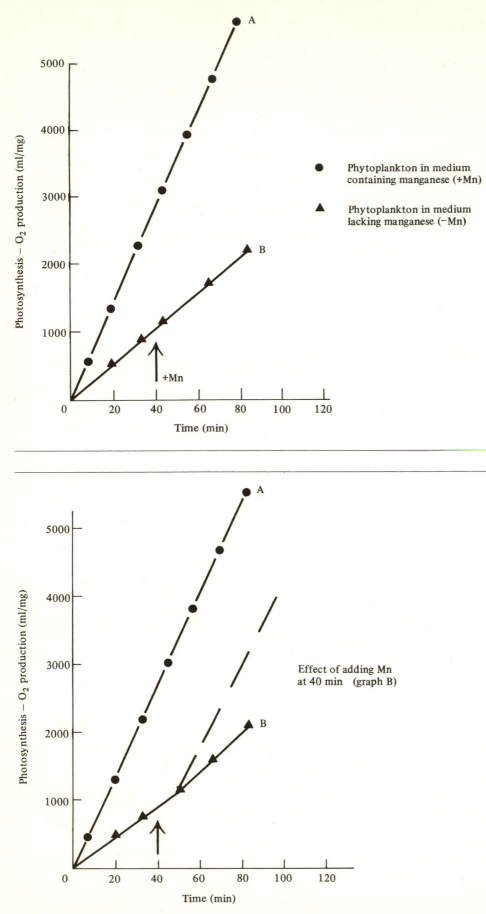

65   You will notice that the addition of Mn does not have an immediate effect on increasing the rate of photosynthesis — why?

_____

_____

It takes time for the Mn to be taken up by the phytoplankton and then taken to the appropriate 'locality' in the cell where it exerts its effect.

66   Again, if Mn is essential for maximum rates of photosynthesis, predict (i) the effect of adding Mn after 30 min to a Mn-deficient culture kept in the dark; (ii) the effect then of turning on the light after 80 min with this same culture. Indicate your predictions on the graph in your response booklets.

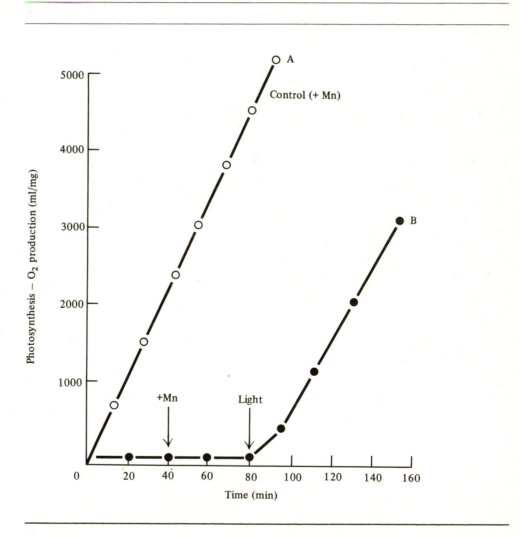

67 Most other elements are essential to living things in one way or another but because many of them are required in such minute quantities, they rarely limit production. They are called the trace elements and include substances like vanadium, boron and copper.

Below is a photograph of cereal plants fed with different quantities of copper. Suggest ways in which man may cause trace element concentration to become limiting on land and in rivers and lakes?

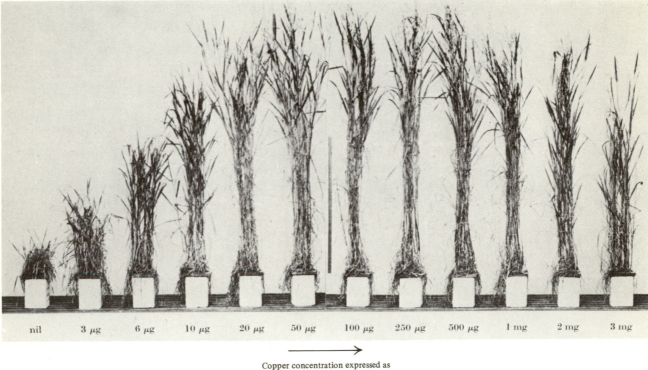

| nil | 3 µg | 6 µg | 10 µg | 20 µg | 50 µg | 100 µg | 250 µg | 500 µg | 1 mg | 2 mg | 3 mg |

Copper concentration expressed as
weights of copper per g nutrient solution

By increasing their concentration to a toxic level through pouring industrial effluent into the rivers and fumes into the atmosphere

68 Let us now draw some general principles from this section.
Which of the following have we found to be true?

(a) Of all the requirements which organisms have of their environment, the one present in amounts most closely approaching the critical minimum will be the limiting one.

(b) Of all the requirements which organisms have of their environment, the one present in amounts most closely approaching the critical maximum will be the limiting one.

(c) a and/or b.

(d) That organisms can produce only in conditions where all of their environmental requirements are optimal, i.e. present in ideal quantities.

(e) That organisms can produce in conditions where none of their environmental requirements are optimal.

c and *e*

69 Make a generalized statement incorporating *c* and *e* above, to describe the controlling effect of the non-living environment in production in phytoplankton.

The presence and success of an organism or group of organisms depends upon complex non-living environmental conditions. Any condition which approaches or exceeds the limits of tolerance of the organism(s) will limit its production, and therefore its success and possibly its presence in that environment. Such controlling conditions are called *limiting factors*.

70 In the context of this interaction, new living material is brought into being and maintained by the processes of photosynthesis and respiration, and a myriad of other chemical processes which go on within the cell. This change from the inorganic to the organic has been summarized for phytoplankton by the following expression:

5 500 000 kJ radiant energy + 106 CO_2 + 90 H_2O + 16 NO_3 + 1 PO_4 + other elements

equals

54 600 kJ potential energy in 3258 g dry wt of protoplasm (living material) comprising 106 C, 180 H, 46 O, 16 N, 1 P, 815 g of other elements, + 154 O_2 + 5 445 400 kJ heat energy dispersed.

What energy difference does the 54 600 kJ potential energy represent?

The energy difference between 3258 g of living material and 3258 g of its simple inorganic 'raw materials', i.e. to convert 3258 g of simple (relatively unordered) molecules into the same weight of complex (more ordered) molecules requires an energy input of 54 600 kJ from sunlight. [Note, however, the amount lost as heat!]

4.2. Interactions between organisms and their living environment. I

4.2.1. Overview

For convenience of working, we have divided this topic into two sections. In this section we shall consider all the organisms in a lake, and examine the exchange of materials and energy, both between different organisms, and between these organisms and their physical environment.

4.2.2. Objectives

At the end of this section you should be able to:
1. Explain, giving examples, the following:
    (i)   the concept of an ecosystem;
    (ii) the trophic interactions between organisms within an ecosystem.
2. Define the following terms:
    autotrophic, heterotrophic, parasite, saprobe, commensal, predation, population, community, photosynthetic efficiency, ecological efficiency, gross production, net production, standing crop, biomass.
3. Demonstrate some skill in the interpretation of reported, graphical and numerical data.

4.2.3. Production and consumption

1   In the previous section we considered only the phytoplankton. Phytoplankton contain the green pigment chlorophyll necessary for photosynthesis, and like all green plants and certain bacteria they are able to synthesize organic compounds (i.e. sugars) from simple inorganic precursors. They are therefore referred to as *autotrophic* organisms. However, there is another group of plankton in the lake which does not contain chlorophyll: the *zooplankton*. From your knowledge of the meaning of the work phytoplankton, suggest the literal meaning of zooplankton.

_____

_____

Floating or drifting animals

_____

2   Suggest one way in which these organisms might obtain their essential sugars, energy, etc.

_____

_____

By consuming living or dead phytoplankton and chemically or physically extracting the sugars, energy, etc. from them.

Note. Any organism which is unable to synthesize organic compounds from simple inorganic precursor molecules is referred to as *heterotrophic.*

---

3   The process whereby zooplankton consume phytoplankton is called *ingestion.* But many of the polysaccharides (polymers of sugars) and proteins (polymers of amino acids) found in phytoplankton are different from those found in zooplankton. What biochemical mechanisms must zooplankton possess if they are to utilize the ingested phytoplankton? (Two mechanisms are necessary.)

---

   1.   Mechanisms to break up the ingested proteins and polysaccharides into their constituent units (amino acids and sugars). This process is called *digestion.*
   2.   Mechanisms to resynthesise the units in the molecular sequences required by the zooplankton. These mechanisms are collectively called *assimilation.*

---

4   **Slide 4** shows a number of single-celled phytoplankton organisms. **Slide 5** shows one such cell some time after it had been ingested by the zooplankton organism shown. What differences do you see in the former?

Only the outer cell wall remains; the remainder has been digested away.

---

5   You will notice a clear space around the remains of the phytoplankton cell. This is called the food vacuole. What do you think the zooplankton secretes into this space?

The chemicals involved in digestion

---

6   Why is it necessary to isolate these chemicals from the rest of the zooplankton cell?

To avoid the cell digesting away its own living material

---

7   Having broken down the polymers of the food organism, the zoo-plankton reabsorbs the products through the membrane surrounding the food vacuole. What properties must this membrane possess?

It must be selectively permeable to the products of digestion.

8   The zooplankton organism in slide 5 is single-celled. What modification would you expect ingestion to impose on the structure of the cell membrane?

There must be a passage through the cell membrane which allows relatively large particles to pass into the cell.

9   Below is a sketch showing these modifications. A cone-shaped canal passes back into the cell, and at its mouth are minute hair-like pro-jections which sweep a current of water continually into the canal. The sketch also shows a food vacuole. What part of the canal is shown by **slide 6**?

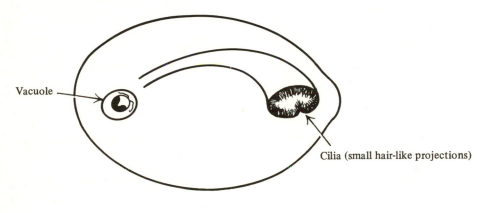

Vacuole

Cilia (small hair-like projections)

The mouth of the canal, with its hair-like projections. The latter are called *cilia*.

10   **Slide 7** shows a zooplankton organism, but this one is multicellular. This organism is the water-flea (*Daphnia*) and it also feeds on phyto-plankton. What shape is its equivalent of the food vacuole?

Tubular. This structure is called the *gut*.

11 **Slide 8** shows another multicellular zooplankton called *Cyclops* which also feeds on phytoplankton. Is the gut of *Cyclops* similar or dissimilar to that of *Daphnia*?

_____

_____

It is similar. The tubular gut of this animal is less easy to see as it feeds primarily on dead phytoplankton which have lost their green colour, as well as living phytoplankton.

12 What term is used to describe intracellular structures such as mito-chondria, chloroplasts, food vacuoles, etc.? (It was mentioned in Book 2, Unit 1.)

_____

_____

Organelles

13 Examine several drops of pond water taken from different places in the beaker. Can you find zooplankton feeding on phytoplankton? Describe them and illustrate their methods of feeding with a sketch.

_____

_____

[Your sketch]
Spend as much time looking at different zooplankton organisms as you find useful or interesting.

14 What feeding alternatives are open to much larger non-green organisms, e.g. fish?

_____

_____

(i)   Feeding on phytoplankton, *and/or*
(ii) feeding on zooplankton, e.g. *Cyclops* and water-fleas

15 This most familiar feeding relationship between organisms is called *predation*, in which one organism ingests, digests and assimilates another. However, there are other mechanisms. Can you suggest what the feeding mechanism of the zooplankton shown in **slide 9** is? This is a single-celled zooplankton and the green cells seen on this slide are living.

_____

_____

The green cells live inside the zooplankton cell, and produce sugars and

40

amino acids, etc. which are absorbed by the zooplankton cell.

---

16 What advantages accrue to the enveloped phytoplankton cells in this
relationship?

They are:
(i)   supported by the zooplankton cell in the light;
(ii) protected from predation.

---

17 The above relationship in which both parties benefit is called *symbiosis*
(common life).
So far we have considered predation and symbiosis as the means
whereby zooplankton obtain the sugars and amino acids they require.
Next we consider an interesting, if bizarre, case in which both predation
and a peculiar form of symbiosis are found.
A minute sea slug feeds on a particular alga by piercing its cells and
sucking out the contents. What type of feeding relationship is this?

Predation

---

18 The gut of this sea slug is not a single tube, but a much-branched pair of tubes, as shown in the diagram, below left. The food is ingested through the orifice marked O, and is eventually passed down to the tips of the branches of the gut. Here, digestive chemicals are secreted onto the food. A cross-section of the cells lining these branched endings reveals both vacuoles and other organelles that one would not expect to find in animal cells. What are they? Examine the micrograph below.

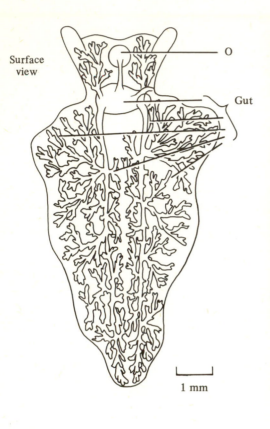

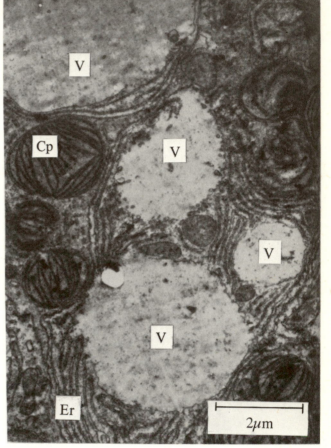

Reproduced by permission of D.L. Taylor, University of Miami

_____

Chloroplasts

_____

19 Since chloroplasts are not indigenous to animal cells, where may they have come from?

_____

From the plant cells, which have been pierced and sucked dry, but not ingested whole.

_____

20 The electron micrograph below reveals their characteristic structure. These chloroplasts appear to be living and undigested. What function may they be performing in the cells?

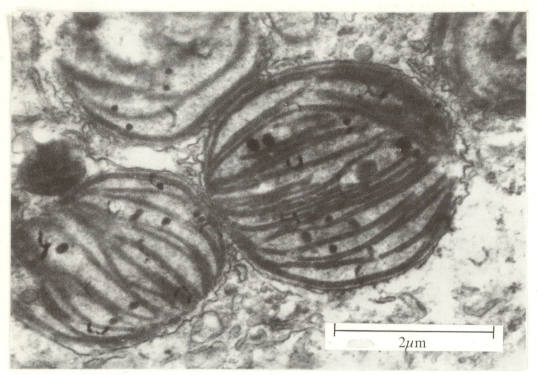

Reproduced by permission of D.L. Taylor, University of Miami. From *J. Mar Biol. Assoc. (UK)*, **48** (1968).

_____

_____

Forming sugars usable by the slug

_____

21 How would you show that photosynthesis is taking place in these chloroplasts?
(*Hint*. the technique has already been introduced in the first section, frame 51.)

_____

_____

Place the slug in an atmosphere containing $^{14}CO_2$, then section and make an autoradiograph (see page 25).

_____

22 Below is just such an autoradiograph. The arrows indicate the position of branch endings of the gut. What does this tell us?

0.1mm

The chloroplasts are actively photosynthesizing in the animal cells.

23 What benefits accrue to the chloroplasts?

They are provided with an environment in which they can function normally.

24 What is the relationship between the slug and the chloroplasts called?

Symbiosis
This situation is rare to the point of being almost unique. Nevertheless, it serves to show how complex and unexpectedly intimate may be the feeding relationship between one organism and another.

25 **Slide 10** shows a stickleback, which is itself a common predator on
    water-fleas and cyclops in lakes. However it is often found to carry an
    unwelcome 'travelling companion'. **Slide 11** shows what emerges if we
    kill the fish and open up its gut. This worm has grown to this size from
    a minute speck, entirely dependent on food resources available within
    the fish. What two main sources of food may it be tapping?

    _____

    _____

    1.   Ingested food, as it passes through the gut
    2.   The living cells of the fish, or their products

26 In fact, the first alternative occurs. This leads to a reduced reproductive
    rate, and general debility and inefficiency of the fish, but not its death.
    How is this relationship different from symbiosis?

    _____

    _____

    Although one partner benefits from the relationship (i.e. the worm),
    the other does not (i.e. the fish).
        This relationship is called *parasitism*. It is not uncommon for one
    parasite to have another parasite feeding on it as well.
    Quote. 'Big fleas have little fleas upon their back to bite 'em,
             and little fleas have smaller fleas and so *ad infinitum.*'

27 So far all the organisms we have looked at have used living plants or
    animals as their source of sugars, amino acids, etc. What other major
    source of these essential substances is available as food to living things
    in the lake?

    _____

    _____

    Dead and decaying material

28 Organisms feeding on this material are described as *saprobic*. An example of a saprobe common in lakes is the freshwater shrimp. One of these is shown below.
Where in the lake would you expect to find saprobes and why?

_____

_____

On the bed of the lake, since most dead organisms sink downwards

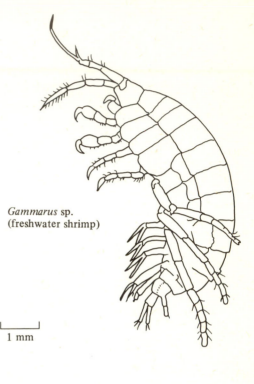

Gammarus sp. (freshwater shrimp)

|��⎯⎯⎯⎯|
1 mm

29 These shrimps rummage around under stones stirring up the layer of sedimented dead material. If one examines the surface of these animals one finds minute unicellular stalked organisms called *Epistylis* (one of which can be seen in **slide 12**). These are also saprobes feeding on the dead materials stirred up by the shrimp. Neither of these organisms harm one another. How does this relationship differ from symbiosis and parasitism?

_____

_____

Only one partner benefits and the other is unharmed. This particular relationship is called *commensalism*. This means 'common table'. The feeding activities of the shrimp, supply the needs of the attached single-celled organism.

30 There remains one very important example of feeding to consider. Bacteria are not photosynthetic and are too small to ingest living animals and plants; in fact, they cannot take in anything larger than particles of molecular size. Can you suggest how they might digest dead material without ingesting it prior to digestion?

_____

_____

By secreting their digestive chemicals into the environment in the vicinity of the food, and absorbing the products.

_____

31 Suggest two inefficient aspects of this feeding mechanism.

_____

_____

(a) Much of the digested food will diffuse away and cannot be
     assimilated.
(b) Once outside the bacterial cell, the chemicals cannot be controlled
     by the bacterium.

32 There are some very important consequences of bacterial digestion, as
    indicated by the following experiment.
    A quantity of different types of phytoplankton were placed in a tank
    containing a suspension of bacteria. The tank was left for about eight
    months in the dark, and during that period changes in the concentrations
    of various chemicals in the water were studied. With reference to the
    graph below what appears to be the side-effect from bacterial digestion
    of the dying phytoplankton?

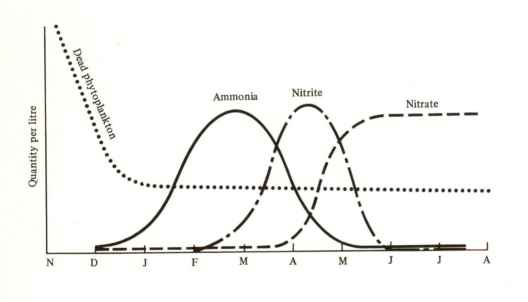

_____

_____

Sequential fluctuations in the proportions of ammonia, nitrite and
nitrate. These fluctuations are caused by different bacteria first
oxidizing ammonia to nitrite and nitrite in turn being oxidized to
nitrate.

33 From what substances in the dead phytoplankton are these chemicals
    derived?

_____

_____

Proteins and amino acids

34 What significance does this release of nitrates into the water have for the maintenance of life in the lake?

_____

_____

Nitrates are essential for plant (including phytoplankton) growth in the lake, and these plants constitute the food, either first hand or second hand, of every other organism in the lake.

---

35 We have thus come full circle. Plants depend on sunlight, CO_2, water and dissolved salts for the production of new living material. Not surprisingly plants are called *producers*. The animals feeding on plants are called *primary* (1°) *consumers*, and the animals which feed on primary consumers are called *secondary* (2°) *consumers*, and so on. All of these eventually die and provide material on which bacteria act. These are called the *decomposers*. The decomposers in their turn reconstitute the inorganic salts, thus closing the cycle. Using the above terminology, how would you describe:

(*a*) a parasite on a secondary consumer;

(*b*) a saprobe?

_____

_____

(*a*) This is a tertiary consumer.

(*b*) This is a decomposer.

---

36 Amongst the producers, primary consumers, secondary consumers and decomposers, which are *autotrophic* organisms and which *heterotrophic*?

_____

_____

Only the producers are *autotrophic*, all the rest are *heterotrophic*.

---

37 By way of summarizing our information on feeding in a pond, a flow diagram has been drawn up below to include some of the organisms we have mentioned. The diagram is divided into horizontal strata called *trophic (feeding) levels*, and is called a *food chain* (when linear) or a *food web* (when more complex). Copy the following food web into your response book and, in place of the circles on the arrows, insert the names of the appropriate feeding relationship, e.g. symbiosis.

     P = Predation
    Pa = Parasitism
     C = Commensalism
     S = Saprobism

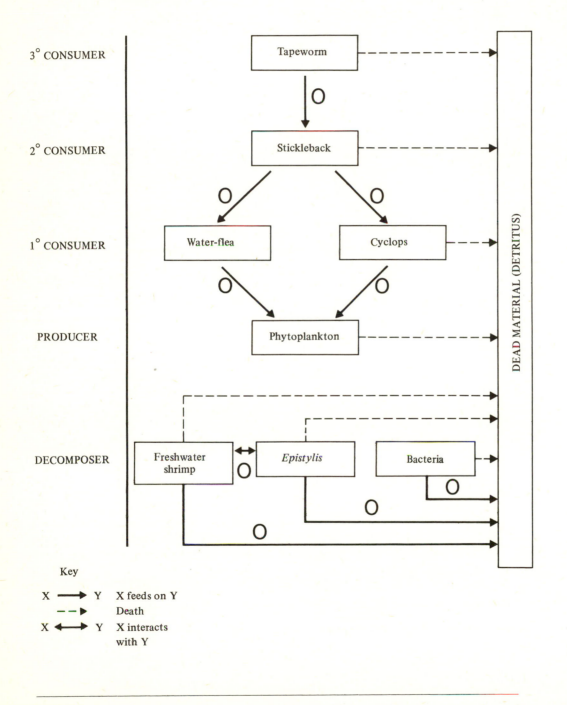

Key

X ——→ Y   X feeds on Y
  – –▶   Death
X ◀——▶ Y   X interacts
            with Y

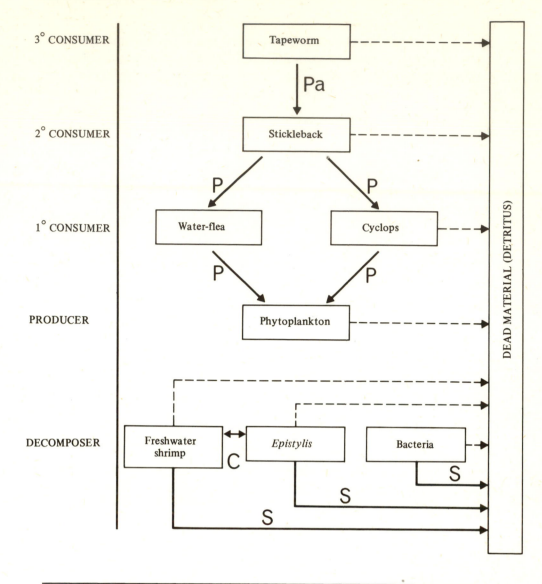

38 Before proceeding, we must define three technical terms; population, community and ecosystem.
A *population* is any identifiable group of the *same species*.
Which of the nine boxes in the food web you have just drawn do *not* refer to populations?

_____

Dead material — because it is a mixture of non-living materials from many species
Phytoplankton and bacteria — because they are composed of more than one species

_____

39 A *community* is any identifiable group of *different species* which interact with one another.
Which of the following are communities?
(a) The non-living components of a lake;

(*b*) the sticklebacks and tapeworms in a lake;
(*c*) organisms having a commensal relationship (commensals);
(*d*) organisms having a symbiotic relationship;
(*e*) the phytoplankton in a lake.

b, c, d and *e*

40 Must a community contain more than one population?

Yes, because a community must contain more than one species and a population cannot contain more than one species.

41 Which of the nine boxes in the food web is not a member of the lake community?

Dead materials

42 At what trophic level do new raw materials enter the lake community?

Producers

43 If production continues, is light or are chemicals likely to be the most limiting factor?

Chemicals, as they will be incorporated into the living organisms.

44 So if we want to study a self-contained unit we need to include chemicals and dead materials as well as the community of living organisms. We call such a unit an *ecosystem* and define it as follows. An *ecosystem* is any identifiable unit in nature consisting of both living organisms and their physico-chemical environment in which there is a continuous exchange of inorganic materials between the living organisms and their environment.

Which trophic levels must be found in all ecosystems? Why?

Producers and decomposers, as they are responsible for incorporating inorganic materials from the environment and returning inorganic materials to the environment.

45 Which of the following are ecosystems?
(*a*)   A lake, its living and non-living components;
(*b*)   the zooplankton in a lake and their decomposers;
(*c*)   a culture of phytoplankton and their decomposers;
(*d*)   a single *Chlorella* cell in a beaker of medium and its decomposers;
(*e*)   all the cyclops in a lake and their decomposers.

a, c and *d*

46 The food web in frame 37 showed a qualitative rather than a quantitative pattern, and only included a portion of the lake community. The table below, however, gives quantitative information about the total membership of the lake community.
(i)   Why are the totals for each trophic level (in bold in the middle column) an unsatisfactory indication of the distribution of living material between levels?
(ii)   What alternative can you suggest?
(*Hint.* Compare, for example, stumpknockers and *Arcella* in the table.)

	Quantities found in the column of water under 1 m^2 of lake surface	
	Total number	Individual dry weight (g)
PRODUCERS		
Sagittaria	206	2.8
Algae	27 X 10^{10}	7 X 10^{-9}
Other large plants	3	14.4
Total producers	**27 X 10^{10}**	
1° CONSUMERS		
Hydrobiidae	6103	0.00005
Oligochaetes	9860	0.0000008
Midges	21107	0.00004
Gammarids	58	0.001
Copepods	162	0.000001
Flatworms	54	0.00001
Paleomonetes	187.6	0.04
Hydroptila	18344	0.00004
Elophila	19740	0.00021
*Arcella	338400	0.000001
Mullet	0.024	238
Pomacea	3.7	1.1
Turtles	0.013	5000
*Stumpknockers	1.0	6.0
Oxytrema	10	0.29
Viviparus	6.37	0.15
Total 1° consumers	**41 X 10^4**	
2° CONSUMERS		
Water-striders	0.4	0.0019
Gurinid beetles	0.13	0.022
Leeches	10.13	0.016
Lucania	6.4	0.06
Gambusia	7.51	0.06
Heterandria	1.73	0.04
Stumpknockers	1.0	6
Sunfishes	0.055	17
Catfish	0.009	286
Notropis and *Hybopsis*	0.12	0.1
Hydra and *Craspedacusta*	24 000	0.00001
Mites	55.4	0.00009
Total 2° consumers	**24 X 10^3**	
3° CONSUMERS		
Bass	0.005	102
Lepisosteus platyrynchus	0.005	209
Lepisosteus osseus	0.0004	760
Total 3° consumers	**0.0104**	
DECOMPOSERS		
Bacteria on plant surfaces	6.7 X 10^{12}	4.5 X 10^{-14}
Bacteria on bottom mud	3.3 X 10^{12}	9.0 X 10^{-14}
Crayfish	11	0.37
Total decomposers	**10 X 10^{12}**	

(i) The total for primary consumers gives equal significance to *Arcella* (each individual having a dry weight of 10^{-6} g) and to a stumpknocker (dry weight 6 g); and this is clearly absurd.

(ii) The total dry weight of all the organisms present at each trophic level.

47   The total dry weight at each trophic level is given in the revised table below.

Quantities found in the column of water under 1 m² of lake surface	
PRODUCERS	Total dry weight
Sagittaria	578
Algae	188
Other large plants	43
Total producers	809
1° CONSUMERS	
Hydrobiidae	0.04
Oligochaetes	0.0002
Midges	0.01
Gammarids	0.06
Copepods	0.00015
Flatworms	0.00005
Paleomonetes	1.7
Hydroptila	0.78
Elophila	0.21
Arcella	0.3
Mullet	5.6
Pomacea	4.0
Turtles	6.7
Stumpknockers	6.3
Oxytrema	2.9
Viviparus	1.0
Total 1° consumers	36.6
2° CONSUMERS	
Water-striders	0.12
Gyrinid beetles	0.0028
Leeches	0.12
Lucania	0.39
Gambusia	0.45
Heterandria	0.07
Stumpknockers	6.3
Sunfishes	1.0
Catfish	2.0
Notropis and Hybopsis	0.02
Hydra and Craspedacusta	0.24
Mites	0.005
Total 2° consumers	10.7

Quantities found in the column of water under 1 m² of lake surface	
3° CONSUMERS	Total dry weight
Bass	0.57
Lepisosteus platyrynchus	0.93
Lepisosteus osseus	0.03
Total 3° consumers	1.53
DECOMPOSERS	
Bacteria on plant surfaces	0.30
Bacteria on bottom mud	0.29
Crayfish	4.0
Total decomposers	4.6

These results can be neatly summarized in visual form by drawing an *ecological pyramid* in which dry weight is represented by width, e.g.

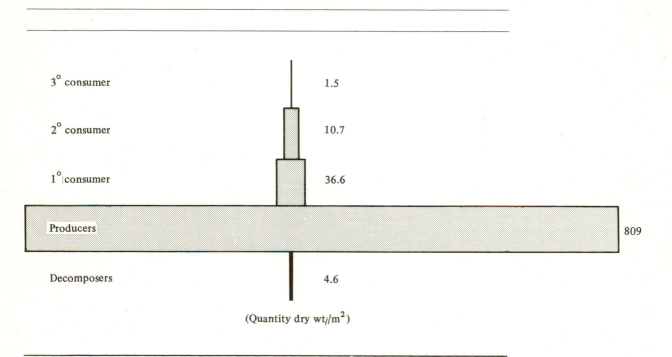

3° consumer

2° consumer

1° consumer

Producers

Decomposers

Quantity

This is only a rough diagram to illustrate the pyramidal shape. Now draw the pyramid again *to scale* to show the distribution of living material between the different trophic levels of the lake community.

3° consumer   1.5

2° consumer   10.7

1° consumer   36.6

Producers   809

Decomposers   4.6

(Quantity dry wt/m^2)

48 Any measurement by weight of living material is referred to as its *biomass*. The quantity of living material (whether measured by weight, size, number, etc.) present at any one time is called the *standing crop*. How would you describe the measurement represented in your pyramid?

Standing crop biomass of each trophic level

49 The 'pyramid' data presented in frame 47 was collected at a given time, i.e. it is an instantaneous measurement. The graph below shows the standing crop biomass of producers and 1° consumers throughout the year. What does this suggest may be the limitation of the frame 47 pyramid?

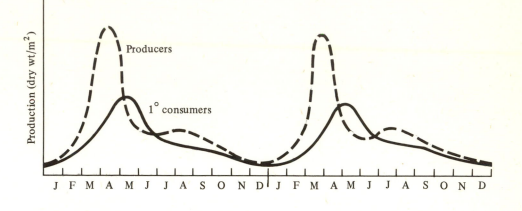

It does not show change with time.

50 What appears to be the period over which this cycle of change occurs?

One year

51 What do the peaks in the frame 49 graph represent?

Periods of massive production

52 The following graph is the same as that in frame 49 but with environmental variables superimposed. Account for the numbered features of the producer curve.

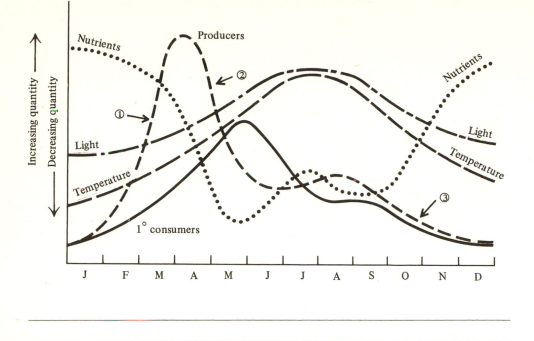

1. Increase in production due to increase in light (and temperature)
2. Decrease in production due to depletion in nutrients (and decrease in standing crop due to predation by 1° consumers)
3. Decrease in production due to decrease in light (and decrease in standing crop due to predation of 1° consumers)

53   The graph in frame 52 also indicates that:
 (a)   The standing crop biomasses of the producers and 1° consumers show little net annual change.
 (b)   The annual production biomass, i.e. the area under the curve, is less for 1° consumers than for producers.
 List three factors that might explain b above.

1. Since there is no net change in the standing crop biomass of the producers, the only available food for the 1° consumers is the net annual production of producers. But some of this production is lost as dead material, so the food available to 1° consumers must be less than the net annual production of producers.
2. Then, even if nearly all the food does get ingested by the primary consumers, not all of it is converted to production. Some of it is converted to energy for heat and movement; and some may not even be digested or assimilated but lost as faeces.
3. The production biomass of 1° consumers is also reduced by predation of 2° consumers.

54   Thus, in order to trace the flow of materials along a food chain, the following quantities must be known for each trophic level.

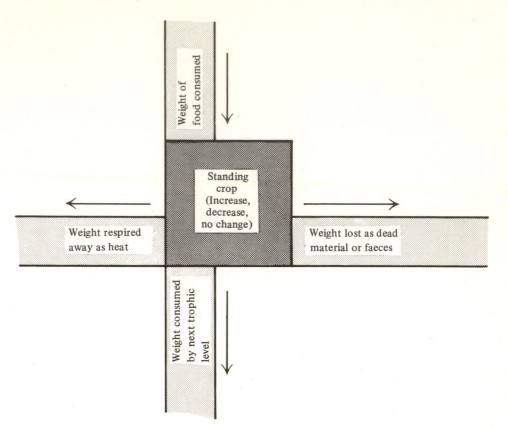

Each of these quantities can be measured in terms of dry weight biomass. Do the following data suggest that the flow of materials along a food chain precisely reflect the flow of energy?

Ensis (a bivalve mollusc)   1 g dry wt has a calorific value of 3500 cal or 14 700 J

Calanus hyperboreus (an animal similar to *Cyclops*)   1 g dry wt has a calorific value of 7400 cal or 31 080 J

No, some organisms have a higher potential energy/g than others.

55   Thus, if we are to trace the flow of energy along a food chain we must find a method for measuring the energy content of living material. This measurement is made by placing a known dry weight of material in an atmosphere of pure oxygen in a calorimeter (a simple calorimeter is shown opposite). The material is burnt, and the heat produced is passed entirely to a known volume of water, whose consequent rise·in temperature is recorded. The potential energy contained in the sample is expressed as calories, where 1 calorie (cal) is equal to the heat required to raise the temperature of one millilitre of water 1 °C. *Note.* In modern scientific literature, kilojoules are used rather than kilocalories to conform with international convention (SI system of units). The conversion from kcal to kJ is simple: 1 kcal = 4.24 kJ.

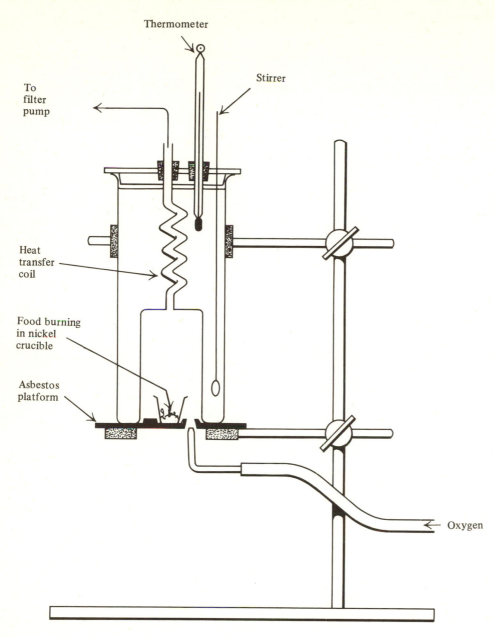

Thermometer

Stirrer

To
filter
pump

Heat
transfer
coil

Food burning
in nickel
crucible

Asbestos
platform

Oxygen

(a) If a sample of phytoplankton weighing 8 g is burnt in a calorimeter, and heats 1 litre of water from a temperature of 10 °C to
98 °C, how much potential energy does the 8 g possess?

(b) What is this potential energy expressed in kilojoules?

(a) 88 × 1000 cal = 88 000 cal or 88 kcal
(b) 370 kJ

56 Below is a diagram showing phytoplankton production in a lake expressed in terms of energy flow. The figures were obtained experimentally as follows. Transparent glass cylinders of known cross-section were placed at random in the lake, with their bottoms plunged in the mud, and their tops projecting from the lake surface. Each cylinder was then filled with lake water, and left for a known period. Production at the end of that period is recorded as the mean kJ 'phytoplankton potential energy' contained in the water beneath 1 m² of lake surface: that is, in this case, kJ/m².

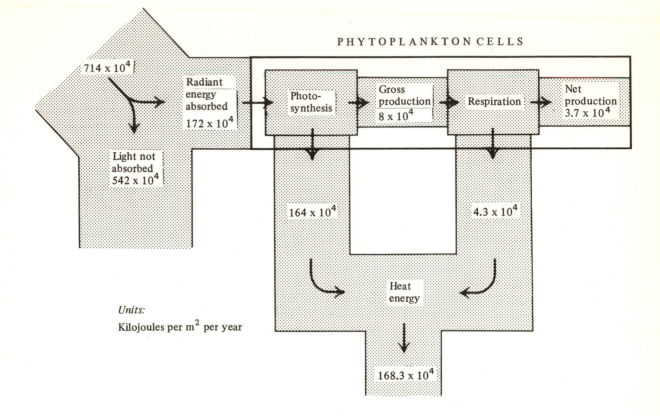

PHYTOPLANKTON CELLS

714 x 10⁴

Radiant energy absorbed 172 x 10⁴

Photo-synthesis

Gross production 8 x 10⁴

Respiration

Net production 3.7 x 10⁴

Light not absorbed 542 x 10⁴

164 x 10⁴

4.3 x 10⁴

Heat energy

Units:
Kilojoules per m² per year

168.3 x 10⁴

What do you understand by gross production and net production?

_____

_____

Gross production is the increase in potential energy as a result of photosynthesis.
Net production is gross production less the energy lost as heat due to respiration.

57 What percentage of the absorbed radiant energy ends up as potential energy in the form of new living material (see frame 56)?

_____

_____

$$\frac{3.7 \times 10^4}{172 \times 10^4} \times 100 = \text{approx. 2\%.}$$ This is the *photosynthetic efficiency*.

60

58   In frame 54, two fates were suggested for the net production biomass
      of producers, 1° consumers and 2° consumers: namely (i) loss as food
      to consumers, and (ii) loss as dead material and faeces (food for
      decomposers). Can you think of an additional fate not hitherto
      suggested?
      (*Hint.* How self-contained is the lake as an ecosystem?)

_____

_____

      Loss by emigration, probably by being carried out of the lake at its
      outflow.
      (*Note.* Heat loss has already been accounted for, otherwise we would be
      using 'gross production' figures.)

═══

59 All the data used so far has been drawn from the work of Odum and his colleagues on the lake Silver Springs during the year 1953-4; and we are now ready to examine the net annual flow of energy through the ecosystem. Our first diagram, shown below, indicates the energy flow through the community defined as 'all living organisms except decomposers'.

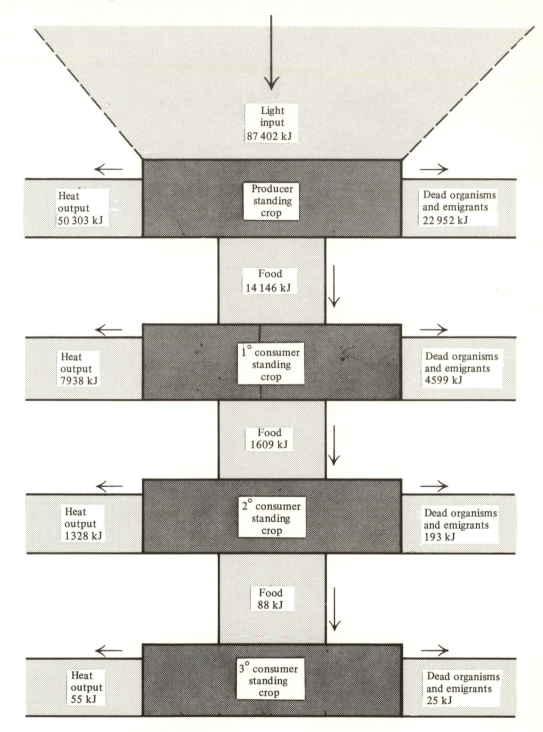

Give the net annual change in standing-crop energy content for each of these four trophic levels.

Producers	0 kJ
1° consumers	0 kJ
2° consumers	0 kJ
3° consumers	+ 8 kJ

60 What is the energy content of the total output of dead organisms and emigrants from the four trophic levels?

27 769 kJ

61 Our second diagram completes the picture by showing the energy flow through the decomposer community and the energy flow in the form of dead organisms and emigrants.

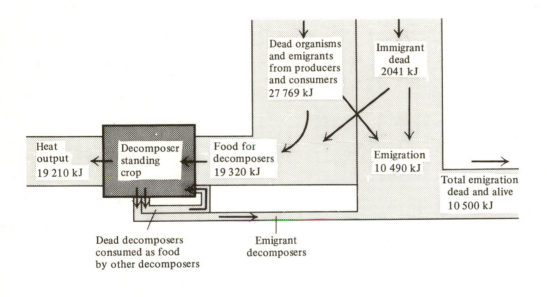

(a) Give the energy content of the emigrant decomposers.
(b) What is the change in the energy content of the decomposer standing crop?

(a) 10 kJ   (b) −10 kJ

62 Your answers to frames 59 and 61 have shown that there is insignificant net annual change in the standing crop.
This being the case, what would you expect to be the *ultimate fate* of the bulk of the energy entering the ecosystem as gross production by the producers?

It will be used to maintain the standing crop, by way of respiration, at each successive trophic level, ultimately being lost as heat.

---

63 What percentage of the energy entering the ecosystem in this way is in fact lost as heat?

$$\frac{78\,834}{87\,402} \times 100 = 90.3\%$$

---

64 We use the term *yield* to describe the amount of production which is available as food without reducing the standing crop. What would happen if more than the yield was consumed as food?

The standing crop would be reduced.

---

65 So how would you define maximum feeding efficiency?

When the food intake of one trophic level is equal to the yield of the level below.

---

66 Consider the following part of a food chain in a stable ecosystem (i.e. one in which the standing crop energy contents at each trophic level remains constant).

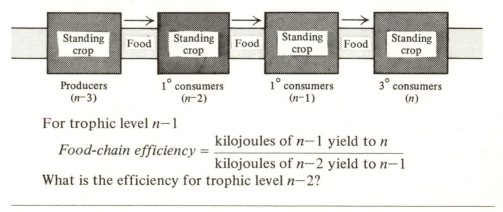

| Producers ($n-3$) | 1° consumers ($n-2$) | 1° consumers ($n-1$) | 3° consumers (n) |

For trophic level $n-1$

$$\textit{Food-chain efficiency} = \frac{\text{kilojoules of } n-1 \text{ yield to } n}{\text{kilojoules of } n-2 \text{ yield to } n-1}$$

What is the efficiency for trophic level $n-2$?

$$\frac{\text{kilojoules of } n-2 \text{ yield to } n-1}{\text{kilojoules of } n-3 \text{ yield to } n-2}$$

---

67   When feeding efficiency is optimal and food intakes are equal to yields (see frames 64, 65) this expression can be written as

$$\frac{\text{kilojoules of } n-2 \text{ yield to } n-1}{\text{kilojoules of } n-3 \text{ consumed by } n-2}$$

We call this ratio the *gross ecological efficiency* and normally express it as a percentage value by multiplying the above equation by 100.
Use the diagram in frame 59 to calculate the gross ecological efficiency for (*a*) 1° consumers, (*b*) 2° consumers.

(*a*) $\dfrac{1609}{14\,146} \times 100 = 11.3\%$

(*b*) $\dfrac{88}{1609} \times 100 = 5.5\%$

Conclusion

In this section we have examined the different feeding interactions present in the lake ecosystem. We have also produced models of the lake community (food chains, food pyramids, flow diagrams), which have enabled us to trace the flow of materials and energy through the ecosystem from sunlight to living material, heat and detritus. This led us to the discovery that the ecosystem we were considering was, in fact, stable in terms of its biomass and energy content. It is now suggested that you read the pamphlet in Section 4.2.4 entitled 'Maintaining life on future voyages into space' as this will draw together most of the ideas we have considered to date.

Also at this point we suggest that you might like to carry out an investigation into the numbers of different kinds of small animals found in leaf litter. For this work you will require the following:

(i)   A selection of leaf litter from different habitats (collect about 2½ kg or 5 lb in plastic bags). It is best to take the leaf litter from below the surface layer (the top is usually too dry for most organisms) together with the top scraping of soil, i.e. about 5 mm.

(ii)   An ordinary light microscope with a low-power objective ($\times 6.3$) or a binocular stereoscopic microscope.
A small petri dish, a fine squirrel-hair brush and some small specimen collecting tubes containing dilute alcohol.

(iii)   A Tüllgren funnel consisting of a funnel with an inverted funnel above as shown in the diagram overpage. If you do not have a metal Tüllgren funnel *large* glass filter funnels painted black on the outside and lined with aluminium foil inside can be used and clamped into position. The top (inverted) funnel should carry a 60 W (or with a large funnel a 100 W) bulb. The lower funnel has a fairly fine sieve onto which samples of the leaf litter are placed.

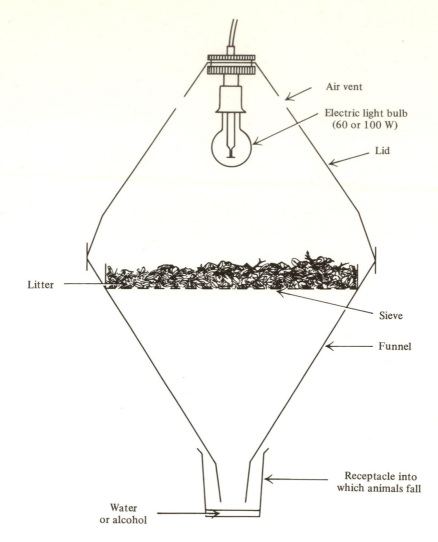

The apparatus works on the principle that most small organisms move away from bright light (i.e. they are negatively phototactic) and very warm and dry conditions. They therefore move to the bottom of the leaf litter and fall through the fine sieve into a collecting beaker (10 ml) containing a small amount of water (or dilute alcohol).

Instructions
Hand sort leaf litter for any larger animals, i.e. obvious to the naked eye, then carefully place samples of leaf litter on the sieve and place in the Tüllgren funnel. Leave for at least one to two hours. Under a low-power binocular microscope identify the main groups of invertebrates as shown in the attached key. Make a count of the numbers in each group, so that you can construct an ecological pyramid of the leaf litter found, as shown overpage.

Locality	Total numbers									
	Isopoda (woodlice)	Acarina (mites)	Diplopoda (millipedes)	Collembola (springtails)	Mollusca (snails)	Chelonethida (pseudoscorpions)	Araneida (spiders)	Opiliones (harvestmen)	Chilopoda (centipedes)	Coleoptera (beetles and larvae)
Beech wood litter	27	6	10	16	14	15	12	1	1	1
Oak wood litter	etc.					etc.				
Pine tree litter	etc.					etc.				
	1° consumer					2° consumer				

The commonest organisms found in most leaf litter localities are the

Collembola	(springtails)
Acarina	(mites)
Isopoda	(woodlice)
Diplopoda	(millipedes)
Chilopoda	(centipedes)
Mollusca	(snails)
Araneida	(spiders)

The Tüllgren funnel will not extract the semi-aquatic animals in leaf litter or soil, such as nematodes and small worms. For this purpose a Baermann funnel is employed, as shown below

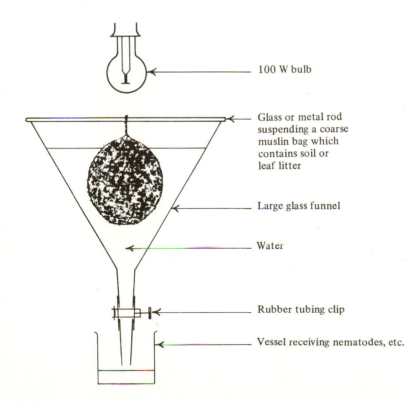

100 W bulb

Glass or metal rod suspending a coarse muslin bag which contains soil or leaf litter

Large glass funnel

Water

Rubber tubing clip

Vessel receiving nematodes, etc.

Some leaf-litter invertebrates*

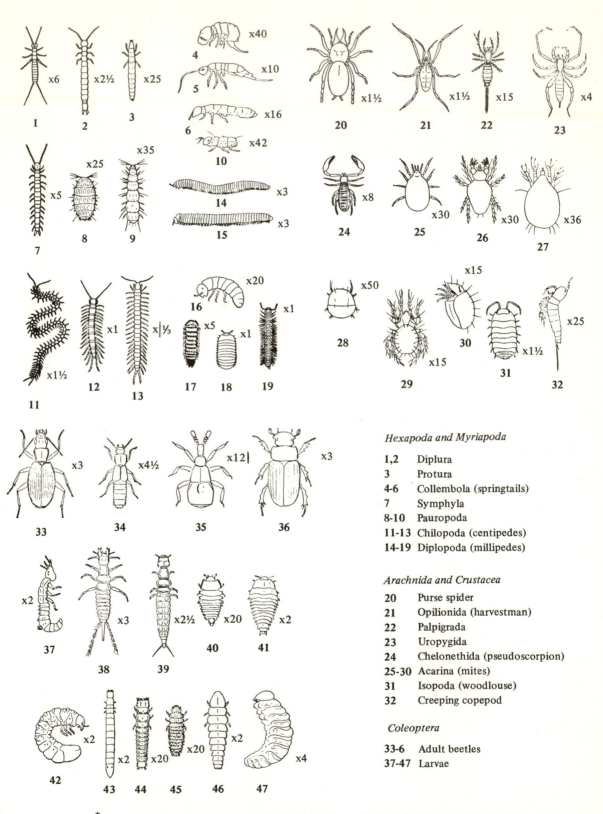

Hexapoda and Myriapoda

1,2	Diplura
3	Protura
4-6	Collembola (springtails)
7	Symphyla
8-10	Pauropoda
11-13	Chilopoda (centipedes)
14-19	Diplopoda (millipedes)

Arachnida and Crustacea

20	Purse spider
21	Opilionida (harvestman)
22	Palpigrada
23	Uropygida
24	Chelonethida (pseudoscorpion)
25-30	Acarina (mites)
31	Isopoda (woodlouse)
32	Creeping copepod

Coleoptera

33-6	Adult beetles
37-47	Larvae

*Reproduced from *Soil Animals*, by D. K. McE. Kevan. H. F. & G. Witherby Ltd., London.

68

4.2.4. Pamphlet: Maintaining life on future voyages into space*

Regardless of when this extended exploration of space by man comes about, the life-support systems for manned spacecraft will have to function for far longer periods than they do now. Our present ones are good for only a few weeks. Systems for the long-term space stations and the trip to Mars will be markedly different from those in the Apollo and Soyuz spacecraft of today.

When we begin to probe beyond the Moon or the nearer planets, we simply cannot utilize a system that operates on the continuous depletion of stores as do the semi-enclosed life-support systems in use today. One alternative, then, is to duplicate in some degree the ecological or closed system of which man is a part on Earth. For the exploration of deep space, man will probably need a microenvironment based upon the exchange of energy between animal and plant and the recycling of water. The heart of such a system is a photosynthetic unit in which man's carbon dioxide is converted into oxygen and food by green plants. Such a system is shown schematically below [Fig. 1].

In speaking of the closed life-support systems, it must be kept in mind that we are thinking only in terms of the matter loop of such a system. The energy loop is quite obviously open, and there appears to be no way of closing it. Thus, while matter can be exchanged continuously in the system, energy for it must be continually supplied from an outside source. The source for our closed system on Earth is the Sun, and so this star suggests itself as the ultimate source for the closed life-support system of the spacecraft — at least for travel within the solar system.

The major units of the ecological system are the energy supply, photosynthetic exchanger, waste-management system, water processor, atmospheric controller, and food-production unit.

Each of these requires several subsystems and the whole must be interconnected by an automatic control system that constantly maintains correct temperature, humidity, and pressure (including partial pressures of several gases). The problems of maintaining such a system in dynamic equilibrium are formidable indeed. To illustrate the complexity of the system, we shall consider only the photosynthetic exchanger.

The most critical element of the entire system is the photosynthetic unit. A colony of algae suspended in water is often suggested for this. At once, we face a problem in deciding which species of alga is best suited for the unit. There are some 40 000 species, but we need not investigate them all because they range in size from the 200 ft long ocean kelps to the microscopic, unicellular varieties that dwell in stagnant ponds. From the standpoint of use in the limited quarters of the spaceship, a very small alga is indicated. Thus, research to date has been confined to the single-celled species such as those shown in the microphotograph below [Fig. 2]. Attention has centred on several of the *Chlorella* (mainly *Chlorella pyrenoidosa*), *Scenedesmus*, *Anacystis*, *Synechocystis*, and *Synechococcus*.

Light for the unit is another critical factor. *Chlorella* typically requires a spectral range of 4 million to 7 million angstrom units for sustained productivity. The conversion of electricity to visible light is only 20 per cent efficient. Furthermore, the maximum efficiency with which algae convert visible light energy to cellular (chemical) energy appears to be between 18 and 22 per cent. The overall efficiency of the algal system seems to be at

---

*This pamphlet was extracted from a book by Mitchell R. Sharpe entitled *Living in Space* published by and © 1969, Aldus Books Ltd, London.

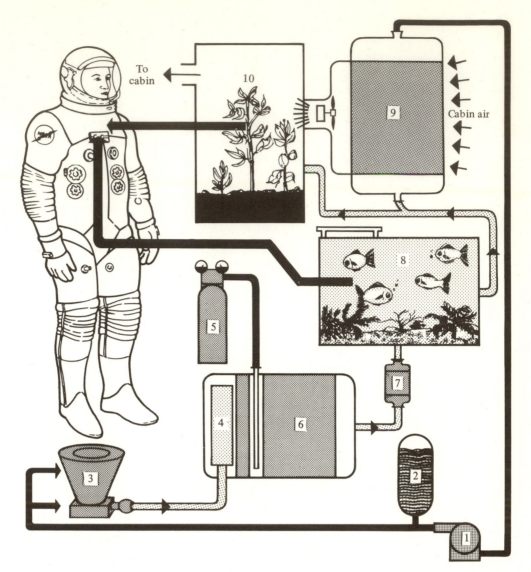

Fig. 1.

A closed ecological system proposed for interplanetary travel. In this system, the pump (1) mixes water from the accumulator (2) with human wastes from the bowl (3); this mixture passes to the comminutor (4), which breaks it down into smaller particles. Oxygen (5) is then introduced into the stream, which passes through redwood bark fibres (6) where bacteria, protozoa, etc. digest fluid. The stream, whose temperature is regulated by a heat exchanger (7), enters a scavenger fish tank (8) where fish eat excess microorganisms not needed for digesting waste. Stream flows next through diffuser membranes (9) which also cleanse water of toxic agents and release carbon dioxide and water vapor. Main stream of water returns to pump; smaller intermittent stream, containing a high concentration of nourishing inorganic compounds, flows to greenhouse (10). Here water vapor, cleansed of bacteria and viruses by a diffuser, passes through to a condenser (not shown), which converts it to storable drinking water. Vegetables convert carbon dioxide to oxygen, which is then returned to cabin. Both fish and vegetables provide food for astronauts.

best of cases, if we consider the conversion of electrical energy to algal cells. Reliable light sources in the spacecraft pose a problem, too. Experimentation shows that, for the purpose, fluorescent forms are superior to incandescent ones, but power requirements are about 4 to 5 kW per man. This need could be met by fuel cells, or solar or atomic sources. Concentrating sunlight by a reflector and diverting it into the photosynthetic exchanger might also be possible.

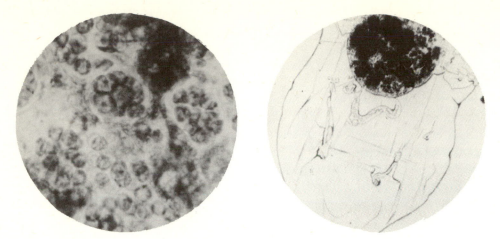

Fig. 2.
Above left: algae, such as this mixture of *Pandorina, Eudorina,* and *Euglenia*
(×325), are often suggested as the green plant for the photosynthetic exchanger in
a closed life-support system for future spacecraft. Right: the water flea, genus
Daphnia, is suggested as an intermediate animal in the system. In theory the flea,
magnified here with stomach full of algae, would eat these tiny plants and in turn
be eaten by the crew. Photos: Dr G.W. Johnston, Mississippi State University.

Perhaps the most critical factor in the closed ecological system is
balancing the respiratory quotient (RQ — ratio of the volume of carbon
dioxide evolved to the volume of oxygen consumed) of the crew with the
assimilatory quotient (AQ — ratio of the volume of carbon dioxide consumed
to the volume of oxygen yielded) of the algae. The balance is critical within
1 per cent. Any deviation leads to a loss of 1 per cent per day of the human
oxygen rechanging the source of the nitrogen supplied for its metabolism, a
regulatory system that could continuously monitor the atmosphere and
maintain the proper RQ/AQ would be complex, to say the least.

Nutrients for the algae are to be supplied by human wastes. The algae, in
turn, become the food for the crew. In addition to the water and carbon
dioxide the algae need for the formation of new cell material, they also
require fixed nitrogen and certain mineral salts. The liquid and solid wastes
of the crew, with some processing, can supply the nutrient requirements of
the algae. If we assume that the algae alone are used to balance the human
respiration, then some 21 oz [500 g] of dried algae per man-day will
accumulate. It is extremely doubtful whether a man could consume this
amount of algae for very long, although they are rich in essential amino acids
(except for sulfur-containing methionine and cysteine) and vitamins, and
consist of 40 to 60 per cent protein, 10 to 20 per cent fat, and 20 per cent
carbohydrate. Experiments do indicate that man can tolerate a diet containing about 3.5 oz of algae but that greater amounts produce a variety of
gastrointestinal disorders.

Food technologists in both the USA and the USSR suggest that the closed
ecological life-support system could have intermediate forms of life as a food
supply for the crew. Variously mentioned are yeast, fungi, mushrooms,
water-fleas [see Fig. 2], fish, snails, slugs, eels, rabbits, chickens, and goats.
Also mentioned are plants such as sweet potato [see Fig. 3] cabbage and
duckweed. The algae would be consumed by the fish or other animals, which
would in turn be eaten by the crew. Such suggestions, however, overlook the
enormous task of recycling feathers, hair, hide, horns, and offal through the
closed system. Dr Robert G. Tischer, a noted microbiologist in the USA,

71

whimsically says that what is needed is a 'dwarf ruminant, probably the size of a cat, which has no hoofs, claws, hair, horns, etc. — in fact [one] which can be eaten in its entirety'.

Other imaginative sources of food for the future space traveller include the production of formaldehyde from methane. This highly toxic compound would then be broken down into 'palatable and useful sugars'. Only a little morbid imagination is needed to suggest a more practical use for formaldehyde aboard the spacecraft. Somewhat more relevant and practical was an experiment in which inmates of an American prison lived for six weeks on a special liquid. It was made of 20 amino acids, fat, several carbohydrates, vitamins, minerals, and water. A cubic foot of the substance is said to be capable of supplying 2000 cal/day for a month. Even more fanciful are suggestions that structural materials within the spacecraft be made of edible compounds, as emergency food supplies. Searching for precedents, proponents of such ideas point out that in Japan beer bottles of powdered and compressed globefish are available.

In view of these difficulties associated with algal systems, research today is also considering more efficient biological systems. The leading candidate suggested as a replacement is the bacterium *Hydrogenomonas eutropha*, which can reduce carbon dioxide and produce cell material and water. The energy for this system comes from the bacterial oxidation of hydrogen with molecular oxygen. In the proposed life-support system, the bacteria function only as a reduction unit for carbon dioxide. Oxygen must be obtained from the electrolysis of water. Such a system is estimated to have an overall efficiency of 30 per cent. However, much work remains to be done on the nutritional value of *Hydrogenomonas* (even though it is 70 per cent protein) and its dietary acceptable by man.

Perhaps the ultimate answer to the problem of providing food in space lies in developing a spacecraft that can travel at velocities close to light. The relativistic effects could solve our nutritional problems. George Gamow points out that at such speeds the astronaut's digestive system would slow down, just as his clock does, by a factor of 70 000. Thus, he would need only one meal for each 6570 eaten by earthbound associates.

Such interesting speculations aside, the preparation of food for use in very long-term space voyages is one of the more stimulating and challenging areas of space research today.

Fig. 3.

Right: edible green plants such as these sweet potatoes are also suggested as sources of oxygen and food for the crew of the future spaceship with a closed ecological life-support system. Photo: Columbus Laboratories, Battelle Memorial Institute, Columbus, Ohio.

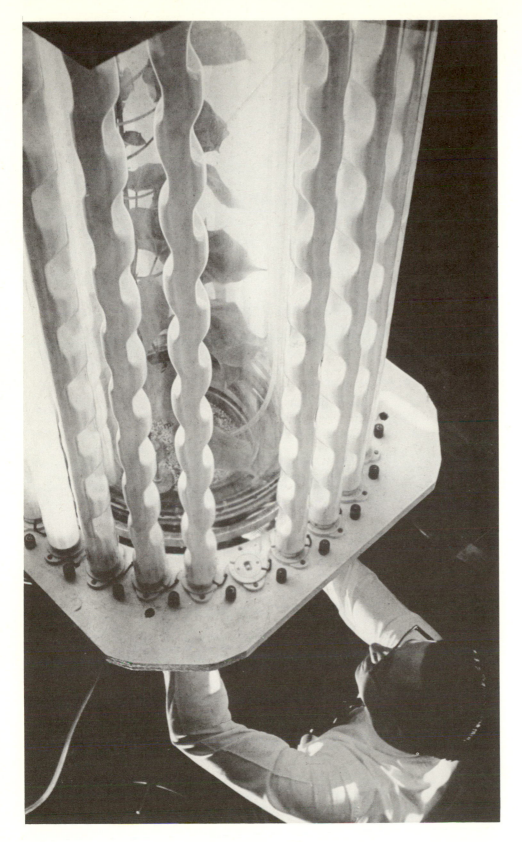

4.3. Interactions between organisms and their living environment. II

4.3.1. Overview

In this section, the way in which the lake or pond changes with time will be examined. We shall study the nature and mechanism of this change with reference to the roles of its living and non-living components.

4.3.2. Objectives

At the end of this section you should be able to
1. Explain, giving examples, the following:
    (i)   the processes of invasion, colonization, competition and succession;
    (ii) the distinctive characteristics of mature as compared with immature ecosystems;
    (iii) the concept of an ecological niche.
2. Demonstrate some skill in the interpretation of reported, graphical and numerical data.

4.3.3. Change in ecosystems

1   The stability of the lake ecosystem we were examining in section 4.2 is indicated by two factors:
    1. The annual change in standing crop biomass was virtually zero at all trophic levels; and
    2. The energy entering the system through photosynthesis (P) was virtually equal to the energy leaving the system through respiration (R).
    Two questions, however, arise.
    Was this ecosystem always stable? And will it always remain stable?
    Explain as far as you can how stability in the ecosystem might be upset.

_____

_____

    Any marked change in the conditions would upset the stability, but we don't know enough about the lake and its surroundings to know whether such changes would be likely every few years or only on a much longer time scale.

_____

2   Let us look, however, at a simple laboratory ecosystem.
    A beaker containing 300 ml of nutrient medium was inoculated with a sample of pond water containing 0.07 mg dry weight of living material. The beaker was sealed and placed in the light. This constituted the ecosystem under consideration.
    Over a period of 100 days measurements were taken and the following results obtained

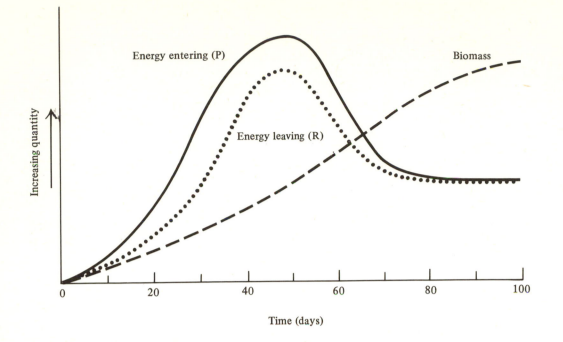

In the last 10 days of the experiment what was happening to:
(a)   the rate of change of biomass;
(b)   the rate of change of (P), the energy entering the system;
(c)   the rate of change of (R), the energy leaving the system;
(d)   the ratio $\dfrac{P}{R}$?

With reference to (a), (b) and (c) the rates of change are all tending to 0.

(d)   The ratio $\dfrac{P}{R}$ has become almost equal to 1

In other words, under the prevailing conditions the laboratory eco-system is becoming stable.

3   We call an ecosystem which has achieved this degree of stability 'mature'; and a system which is still changing 'immature' (i.e. our laboratory system during its first 70 days).
How do biomass and net production vary between the immature and mature states of the ecosystem in the graph?

	Immature	Mature
Biomass	Low	High
Net production	High	Low

4   Let us now summarize what we know about immature and mature ecosystems.

Draw up a *copy* of the following table in your answer books, and insert the appropriate description —chosen from the alternatives given under each of the headings in column II.

I	II	Immature	Mature
Ecosystem energetics	$\frac{P}{R}$ (>1 or 1)		
	Net production (high or low)		
Ecosystem structure	Biomass (high or low)		

I	II	Immature	Mature
Ecosystem energetics	$\frac{P}{R}$	>1	1
	Net production	High	Low
Ecosystem structure	Biomass	Low	High

5   So far we have considered only the quantitative changes which occur in an ecosystem with time. Let us now look at the qualitative changes which occur.

The results of two similar ecosystem experiments as outlined in frame 2 have been combined in the results given below. The bacteria listed are represented by a number, and in the case of the first a record of both the vegetative and spore phases of its life cycle have been recorded.

At the time of innoculation (day 0), only two types of bacteria in their vegetative form were seen to be abundant, although another was common. There were some signs of unicellular algae. No other records were taken on this day.

Qualitative observations made on the following days after addition of innoculate

	Days													
	0	1	5	10	20	30	40	50	60	70	80	90	100	
Bacteria 1 (veg.)	A	A	A	–	–	–	–	–	–	–			–	DECOMPOSERS — Unicellular bacteria
(spores)	–	C	R	R	R	C	C	R	R	R			R	
Bacteria 2	A	A	A	–	–	C	–	C	C	C			R	
3	C	A	C	C	C	C	A	A	A	C			C	
4	C	A	A	A	A	A	A	A	A	A			A	
5	C	C	C	A	A	A	A	A	A	R			A	
6	C	A	A	C	C	–	A	A	A	A			A	
7	C	–	–	–	–	–	C	C	A	A			C	
8	A	–	–	–	C	C	C	C	R	A			A	
9	R	R	R	R	R	R	R	C	C	R			–	
10	R	R	R	R	R	R	R	C	C	A			R	
Chlorella	R	C	A	A	A	A	A	C	C	C	R	–	–	PRODUCERS — Unicellular algae
Schizothrix	R	C	C	C	A	A	A	A	A	A	C	C	R	
Oscillatoria	–	–	–	–	–	R	R	C	C	A	A	C	C	PRODUCERS — Filamentous algae
Lyngbya	–	–	–	–	–	–	R	R	C	A	A	A	C	
Mongeotia	–	–	–	–	–	–	–	–	–	R	C	A	A	
Spirogyra	–	–	–	–	–	–	–	–	–	–	C	C	A	
Daphnia	R	C	C	C	A	A	A	R	–	–	–	–	–	1° CONSUMERS – GRAZERS — Multicellular animals
Copepods	–	–	–	–	R	R	C	C	A	A	A	C	C	1° CONSUMERS – DETRITUS FEEDERS
Ostracods	–	–	–	–	–	R	C	C	A	A	A	A	A	
Nematodes (worms)	–	–	–	–	–	–	–	–	R	–	C	C	A	

Key
A = Abundant
C = Common
R = Rare
– = Absent
? = Not known

Characterize the mature and immature stages of this ecosystem with respect to:

(*a*) organism size (large/small);

(*b*) organism complexity (simple (unicellular)/complex (multicellular))

	Immature	Mature
Organism size	Small	Large
Organism complexity	Simple	Complex

6   Another qualitative feature of interest is indicated in the diagram below.

KEY

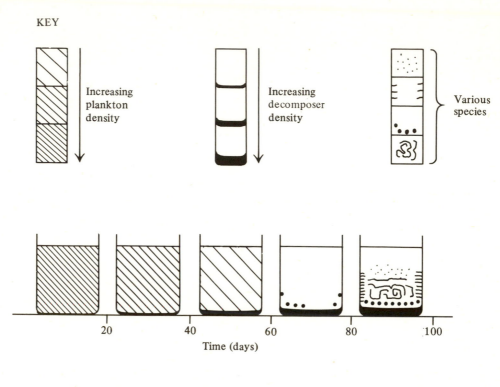

Time (days)

In the immature ecosystem the level of structural organization is _____(high/low), whereas in the mature ecosystem the level of structural organization is_____(high/low).

Low; high

7   The following information was also provided by the laboratory
     ecosystem described in frame 2 (page 75).

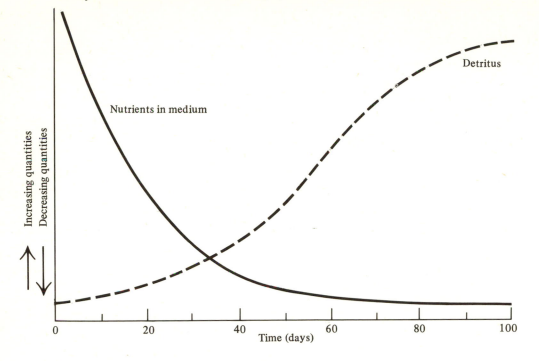

(a)   What general qualitative changes are apparent in the primary
       consumer population with increase in age of the ecosystem?
(b)   What happens to the nutrients which are disappearing from the
       medium as the ecosystem ages?
(c)   How does the importance of detritus as a source of nutrients
       change with the age of the ecosystem?

(a)   There is a move from grazers to detritus feeders, with appropriate
       changes in food webs.
(b)   The nutrients are locked in the biomass. Thus the nutrient supply
       comes under the control of the ecosystem; the mineral cycles are
       closed.
(c)   Detritus, as a result of terminal bacterial decomposition becomes
       the main source of nutrients in the latter stages of ecosystem
       development.

8   We have seen that the structural organization of the mature ecosystem
     is high (i.e. more complex) with a greater variety of organisms and a
     greater species diversity. (Compare day 1 with day 70 in the table
     shown in frame 5). Would such a situation give rise to a smaller or
     greater degree of stratification (vertical layering) in our laboratory
     ecosystem?
     (*Hint.* Consider the distribution of organisms at various depths in the
     beaker.)

A greater degree of stratification in the *mature* ecosystem

---

9 The structural organization of the ecosystem can only be changed by the individual species present. Yet these appear to vary with time as one 'stage' succeeds another.

Since the environment was not interfered with from outside, how do you account for this succession of individual species?

---

1. The spores, eggs or adults of all species found as adults at any time in the ecosystem must have been present at the beginning of the ecosystem.
2. Environmental conditions at the beginning must have suited the earlier organisms.
3. These organisms must have changed the environment in such a way as to make it less favourable for themselves and more favourable for their successors.

This process is called a *succession* and it tends towards a mature ecosystem with properties such as we have described.

---

10 The technique of examining the colonization and succession of organisms on clean surfaces has received considerable attention because of its practical importance in the 'fouling' of ship bottoms and piers by barnacles and other marine organisms. Let us therefore examine in detail the stages of succession in a fresh-water pond using this technique. Twenty small areas were selected at random at the bottom of a shallow fresh water pond (i.e. a mature freshwater ecosystem). These selected areas were first cleared of living organisms and then a clean glass slide (the surface) was placed on each of the cleared areas. The individuals appearing on each slide over a given time were counted and identified. The change in the number of individuals and the number of different species occupying each slide was recorded. The results are given in the graphs below; and each of the twenty dots indicates the number of organisms and the number of species obtained on a single slide.

(*Note.* The organisms primarily involved in this study are bacteria, plankton and other small organisms, but it is *not* necessary for you to know the species names. The ecological principles involved are of much greater importance.)

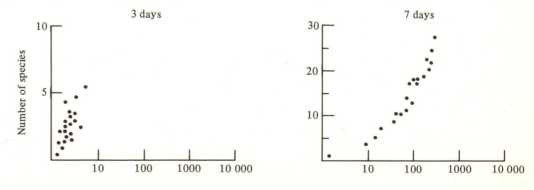

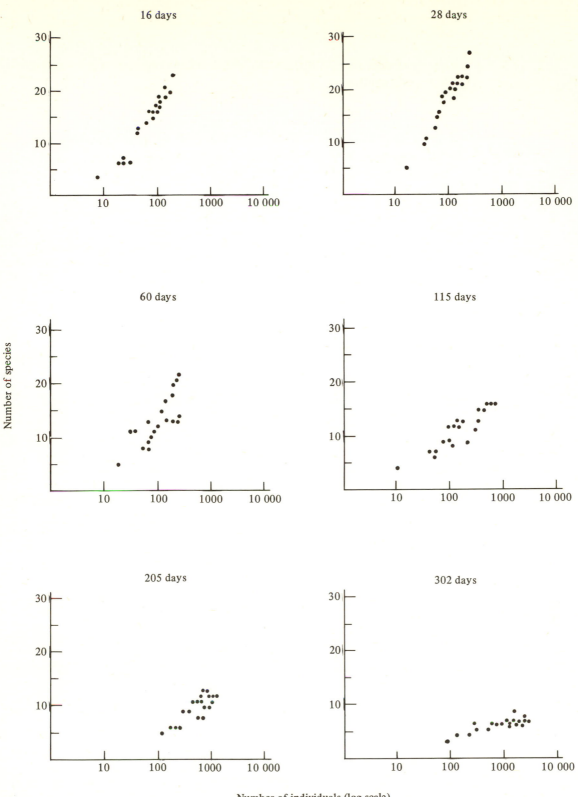

Number of individuals (log scale)

Where do you think the initial individuals come from?

_____

_____

The surrounding areas. This stage is called *invasion*. Clearly our
laboratory ecosystem was unlike 'natural' ecosystems in that invasion
was impossible.

11   There are many more species that invade than ever settle and multiply on the slides. Why is this?

_____

_____

The conditions of this environment are unsuitable for them.
The species which actually settle are called colonizers.
This is the *colonization* stage.

12   Does the number of species increase/decrease/stabilize as time goes on?

_____

_____

It decreases.

13   Why do you think this is?
(*Hint.* Look at the change in number of individuals.)

_____

_____

Because some factor (in this case space) has become limiting and different species are competing with one another for the available resources. Those species most ideally adapted to this particular environment will succeed at the expense of the others. This is called the *competition* stage.

14   Would you expect the early *colonizers* of the glass slide to be producers, decomposers, 1° consumers or 2° consumers?

_____

_____

Producers

15   Which of the following would you expect to be the first *competitors* for the glass slide?
        Producers with producers
        Producers with 1° consumers
        Producers with decomposers

_____

_____

Producers with producers

16   We observed earlier (frame 8) the process of succession in a laboratory ecosystem and also in a natural freshwater system (frame 10). We saw

that this process in our laboratory ecosystem was made possible by virtue of changes brought about in the system by successive components of the succession, with the physical environment (except for light) apparently playing no role in the succession (i.e. the beaker was sealed and placed in the light).

In the pond experiment (frame 10) on the other hand, the overriding limiting factor was *shortage of space* on the slide surface.

Now consider a second experiment in which slides were placed at the bottom of the pond as before, but this time the areas were *selected* according to whether they were rich in nutrients (i.e. nitrates, phosphates, etc.) or poor in nutrients. The results are shown below, where again each dot or dash represents the measurement obtained from one slide.

(a) What difference do you notice between succession in the nutrient-rich environment and the nutrient-poor environment?

(b) In what way does this suggest that the physical environment may affect a succession?

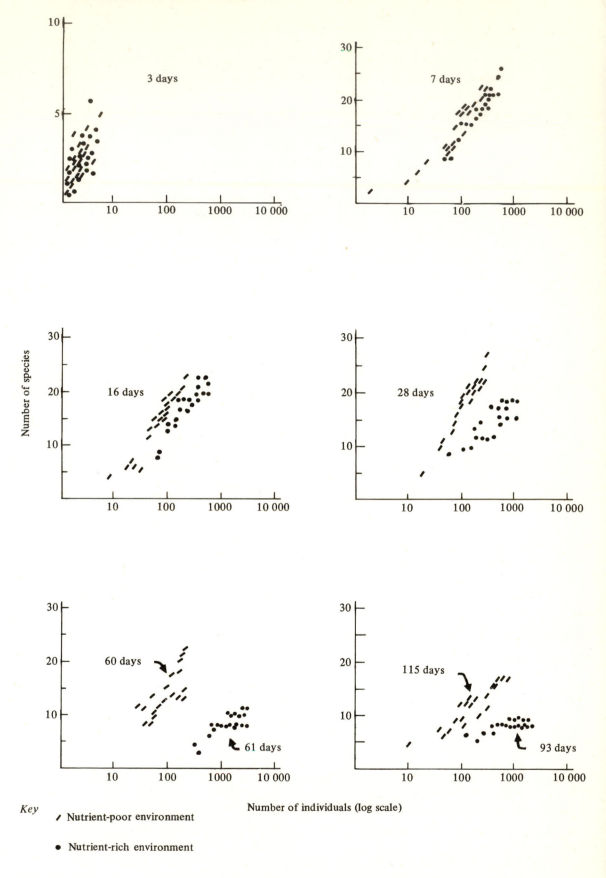

Key

Nutrient-poor environment

Nutrient-rich environment

Number of individuals (log scale)

(a) The process of succession appears to take place at a faster rate in the rich environment.

(b) The organisms respond to a richer environment by faster growth, and the quicker organisms grow, the more rapidly space becomes limiting. Thus the physical environment can affect the *rate* at which an ecosystem matures.

17 If succession on each slide has followed the same pattern as our laboratory ecosystem, would you expect some of the species present at 93 days in the rich environment to be *different* from those present at day 7?
Give your reasons why.

Yes. Because the early colonizers would have changed the nature of the environment, permitting the invasion, colonization and successful competition of other species. Thus we see that invasion, colonization and competition are continuing features of a succession.

18 However, the ten or so species found on each slide at 93 days do not even approach numerically the vast number of species found on the bed of the lake as a whole.
What property of these two environments, one large the other small, do you think are responsible for this difference in species number?

The small environment has little variation, and the large environment has much variation. Thus many different species can find conditions in which they can compete successfully in the large environment, whereas on a glass slide, only relatively few species can find conditions in which they can compete successfully.

19 We shall now consider another ecosystem, one which illustrates a second effect of the physical environment on a succession.
The map on the next page shows the drainage ditches of the flood plain of a tidal river. The preceding pioneer stages of succession in these ditches were not studied, but the subsequent stages were observed. Ditch B, further inland, receives only fresh water; whereas ditch A nearer to the sea receives brackish water (a mixture of fresh and sea water) seeping from the river as it ebbs and flows with the tide.

20 **Slides 13 and 14** show ditches A and B on the map. What do you notice about the surface of these ditches?

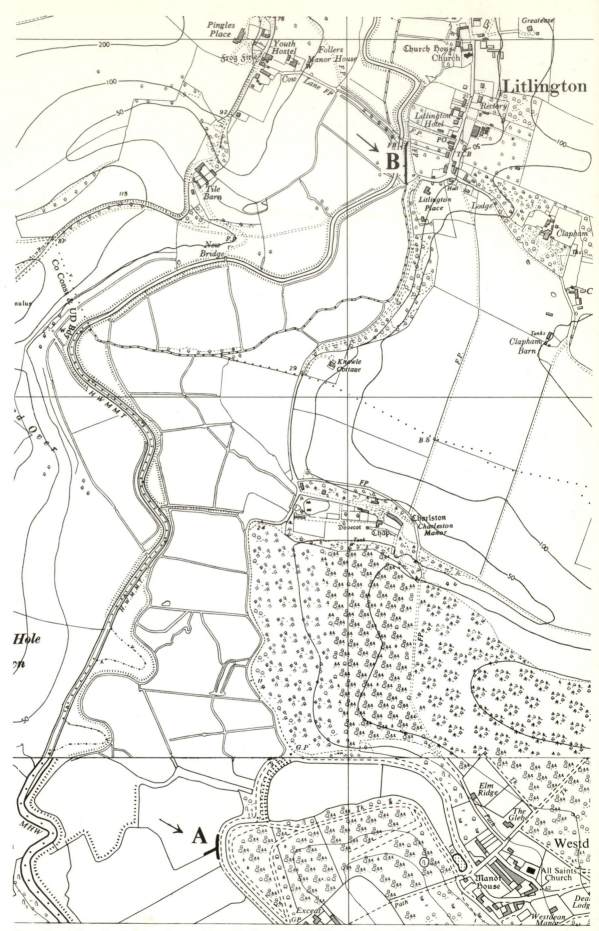

Reproduced from the Ordnance Survey Map with the sanction of the Controller of Her Majesty's Stationery Office, Crown Copyright reserved.

They are entirely covered with vegetation.

---

21 **Slide 15** is a close-up of a representative part of the vegetation on the surface of ditch B.
Is the vegetation homogeneous (one species only) in this case as far as you can see?

Yes

---

22 What possible explanations would account for this homogeneity?

(*a*) The single species dominating this ditch is the final successful competitor of a particular successional stage.
(*b*) This species was the only colonizer of the niche.

---

23 **Slides 16 and 17** are close-ups of the plants on the surfaces of ditches A and B respectively. Clearly they are different. How do you account for this difference, given that the ditches are geographically near to one another?

The nature of the species *colonizing* the two environments was determined by the nature of the two environments, i.e. saline in the case of ditch A and freshwater in the case of ditch B. The eventual successful competitors (if there was more than one colonizer) were different in each case.

---

24 If we regard the flood plain as the ecosystem, what controlling influence does the physical environment exercise?

It determines the pattern of vegetation (and other organisms) at any given time.

---

25 The vegetation on the surface of ditch B is *not* homogeneous. Starwort certainly covers most of the surface, but other species are also found. Thus the distribution of plants occupying the surface of the water is shown overpage.

Frogbit
(floating)   Blanket
weed   Water forget-
me-not (emergent)   Starwort   Duckweed
(floating)   Starwort
(emergent)

Vertical transect diagram showing floating and emergent plants

Surface coverage of a 1m transect across the ditch

1m

Key

 Starwort

 Water
forget-me-not

Blanket
weed

Duckweed

The appearance of the plants is shown in the slides listed.

Slide 18   Starwort
Slide 19   Water forget-me-not
Slide 20   Blanket weed
Slide 21   Duckweed
Slide 22   Frogbit

If, eventually, starwort completely covers the ditch surface, which of the following could we *definitely* say about starwort in relation to its living and non-living environment? Think carefully.

(*a*)   At no point in the ditch were conditions limiting on starwort production.

(*b*)   At all points in the ditch conditions were optimal for starwort production.

(*c*)   All the other plants apart from starwort represented an earlier stage in the succession.

(*d*)   The complex of environmental factors at every point in the ditch favoured starwort for the surface of the ditch, rather than the other plants.

(*e*)   Starwort could tolerate a wider variation in each and every environmental factor than the other plants.

(*f*)   At some points in the ditch, conditions were optimal for starwort production.

(*d*)

26   If starwort were to cover the whole surface, we can say that conditions in the ditch *selected* starwort in preference to its competitors. Or to put

it another way, starwort was suited or *adapted* to conditions obtaining in the ditch better than its competitors. Adaptation on the part of organisms and selection on the part of the environment are to be discussed in another course. [Joint Universities Genetics Course — Open University Publications].

In fact, starwort never quite makes it. Community-induced changes in the ecosystem promote colonization by species of subsequent stages of the succession, such as water forget-me-not. Thus the *floating* and *emergent* stages of a freshwater succession are present simultaneously, though the latter stage is represented but sparsely as yet. The following diagram shows the difference between free-floating and emergent plants. But as time passes, the environment changes and the 'balance of power' alters. Emergent species begin to predominate. **Slides 23 and 24** show two other common emergents in these ditches, marestails and fool's watercress respectively. With the passage of time, dead material and silt accumulate, raising the level of the ditch bed and leading to massive colonization by emergent species. These in turn cause changes in the environment, amongst which is a significant lowering of the pH. Why do you think this is? (*Hint.* Emergent plants have their non-photosynthesizing roots below the water, while their shoots and green leaves are above the water level.)

Although the roots cannot photosynthesize they can respire, removing O_2 and giving out CO_2. As the CO_2 concentration rises in the stagnant water the pH is lowered.

$$(CO_2 + H_2O \rightleftharpoons H_2CO_3 \rightleftharpoons H^+ + HCO_3^-)$$

27 These emergent species in turn compete and eventually change the environment sufficiently to permit the colonization of the next stage in succession. This is illustrated by **slides 25 and 26**. At first the dominant species is the flag. This in turn gives way to the reed stage of the succession as shown in **slide 25**. As the succession goes on the water becomes shallower and shallower as the ditch fills up with eroded soil (silt) and dead and decomposing plant and animal material (detritus). Eventually as the ditch bed rises above the water table of the surrounding land, rushes take over as seen in **slide 26**, and finally the surrounding vegetation encroaches, in this case, grassland. **Slide 27** shows some typical plants which are often found associated with bulrushes (background); they include great hairy willow herb (purple flower), yellow loosestrife and meadowsweet (foreground).

To recap, two very significant changes are brought about by successive organisms in a ditch succession.

(1)   The accumulation of_____?_____and_____?_____, lead to
(2)   the consequent raising of the level of the ditch bed.

silt, detritus

Succession in a terrestrial ecosystem

Each successive stage in the succession will contain some or all the species from subsequent stages

90

28   In the case of the river, the physical environment (the topography of
the land and the climate) is arresting the succession at an early stage.
Here we meet the third and last way in which the environment is
implicated in succession, i.e. it may set limits which determine how far
succession can go. The question now arises as to whether the dis-
appearance of the ditch means the end of the succession. **Slide 28** was
taken at the perimeter of a deciduous woodland. Shrubs, bushes and
climbers are shown in the foreground; trees in the background. **Slide 29**
is a general view of the woodland with trees of various sizes, and **slide
30** is a close-up of one of them. If the conclusions drawn from our
model laboratory ecosystem apply to all ecosystems, would you expect
grassland to represent the mature stage of our succession or forest?
Give your reasons.

Forest, which has larger biomass per square metre and organisms of
larger size.

29   The succession from grassland to forest is shown in the diagram on the
opposite page. Note that species which predominate at any one time in
a succession are called *dominants*. We should also add that our
description of succession is much simplified. Frequently complex
factors and zonation patterns complicate and limit the extent to which
succession takes place (see frame 28).

30   What other feature of our mature laboratory ecosystem is displayed by
the mature forest?

A high level of structural organization, i.e. *stratification*.

31   Stratification in forests can be more clearly recognized than in our
pond ecosystem. To quote Odum (1971) 'In a forest, the two basic
layers, the *autotrophic* and *heterotrophic* strata — that are characteristic
of all *communities* are frequently distinctly stratified into additional
layers. Thus, the vegetation may exhibit herb, shrub, understory tree,
and overstory tree layers, while the soil (itself) is (also) strongly
layered.'
Define the terms *autotrophic, heterotrophic* and *communities* in the
above context. (*Hint.* All these terms were defined in the previous
section of this book.)

Autotrophic	Organisms in this stratum are capable of synthesizing organic compounds from simple inorganic precursors (e.g. green plants and certain bacteria).
Heterotrophic	Organisms in this stratum are unable to synthesize organic compounds from simple precursors (e.g. animals, fungi, many bacteria).
Communities	Any identifiable group(s) of different species which interact with one another.

32 On the opposite page is an energy-flow diagram for a defined area of a terrestrial ecosystem, i.e. it is towards the end of the mixed woodland stage (see frame 29).
What is the energy entering the system through photosynthesis (P)?

6090 + 2940 + 2100 = 11 130 kJ
This equals heat loss (8820) + increase in standing crop biomass (2100) + small quantity of undecomposed humus (210).

33 What is the P/R ratio, and what does this indicate about the maturity of the ecosystem?

$$\frac{P}{R} = \frac{11\,130}{8820} = 1.26$$

The system is approaching maturity.

34 In what form do you think the gain in producer standing-crop biomass is stored?

Wood

35 Which of the two food chains (webs) shown is more important in terms of energy flow?

Detritus

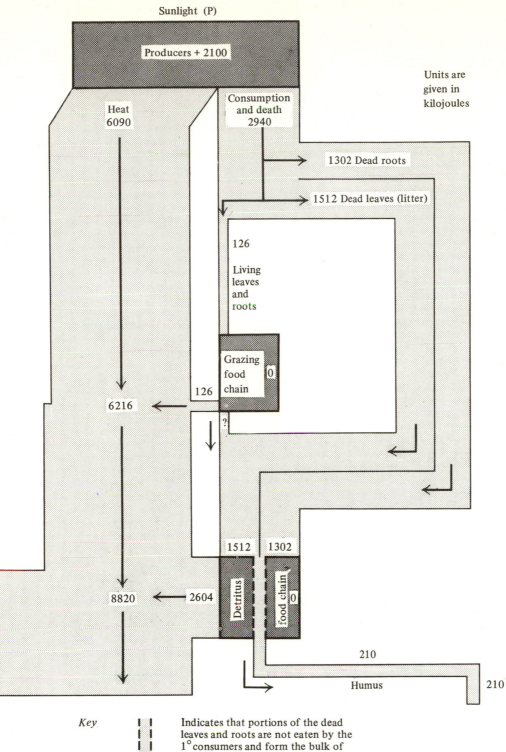

Sunlight (P)

Producers + 2100

Units are given in kilojoules

Heat 6090

Consumption and death 2940

1302 Dead roots

1512 Dead leaves (litter)

126

Living leaves and roots

Grazing food chain

126

6216

8820 ← 2604

1512 1302

Detritus

food chain

210

Humus

210

Key

Indicates that portions of the dead leaves and roots are not eaten by the 1° consumers and form the bulk of the humus remaining at the end of the year.

2 Food-chain standing crop or producer standing crop. Figures indicate annual change in potential energy.

2 → Energy flow

36 Which food chain has the greatest standing-crop biomass?

Detritus, since greater amounts of respiratory energy are expended to maintain it (2604 as against 126 kJ).

37 In what two main forms will the forest be richer in terms of potential energy gain in the succeeding year?

Producer potential energy and humus potential energy

38 The mature forest represents the *climax* of the terrestrial succession. No other species can oust the dominant species and so initiate a further successional stage. After several hundred years, the climax dominant may constitute a pure stand (i.e. large trees predominate).
The nature of this dominant species will differ according to the physical environment, as did the vegetation in our freshwater and brackish ditches. Thus we find that in Scotland, pine is the dominant of the climax forests. In England, where the climate is warmer, its place is taken by oak or beech, the latter being more common on basic soils.
Slide 31 shows an aerial view of a tropical rain forest and again a different species dominates the climax.

39 An organism's status in an ecosystem can be described in two main ways.
   1. In terms of its physical position in the ecosystem: *where* it lives. This is known as its *spatial niche.*
   2. In terms of its functional position in the ecosystem: *how* it lives. This is known as its *trophic niche* (e.g. what it feeds on and when).
   These two descriptions together constitute a definition of the *ecological niche* of an organism.
   Bearing these definitions in mind, consider the following:
   (*a*) A blackbird and a thrush are found in the same garden. Both feed on worms.
   (*b*) In the same garden are found swallows, woodpeckers, and nightjars, all feeding on insects.
   Which of the birds occupy the same *spatial niche*, and which occupy the same *trophic niche*?

All occupy the same spatial niche, but the birds in (*a*) occupy one trophic niche, and those in (*b*) occupy another.

40 How would you modify your answer to frame 39 in the light of the following:

Bird species	Feeding time	Feeding place
Nightjar	Night	In mid-air
Swallow	Day	In mid-air
Woodpecker	Day	In bark of tree

These three birds occupy different *spatial niches* since they feed at different times and in different places.

41 The precise description of the *trophic niche* of an organism may not always be as straightforward as might be expected. The rook is a bird considered by farmers as a pest. When its crop (the part of its gut where its food is stored prior to digestion) is opened, it is found to contain numerous grains of wheat, barley, rye, etc. However, when the crop of a fledgling rook is opened it is found to contain dozens of green wriggling caterpillars. What additional knowledge does this suggest that we need if we are to precisely define the trophic niche of an organism?

A knowledge of its life history and the changes in behaviour and diet from birth to death.

42 In the diagram below the background squares represent environmental factors (temperature, food sources, space, etc.) projected onto one plane. The irregular polygons enclose a set of factors which represent the ecological niches of species A, B, C and D.
Redraw the diagram and shade the areas where competition will occur. Which species is in danger of elimination?

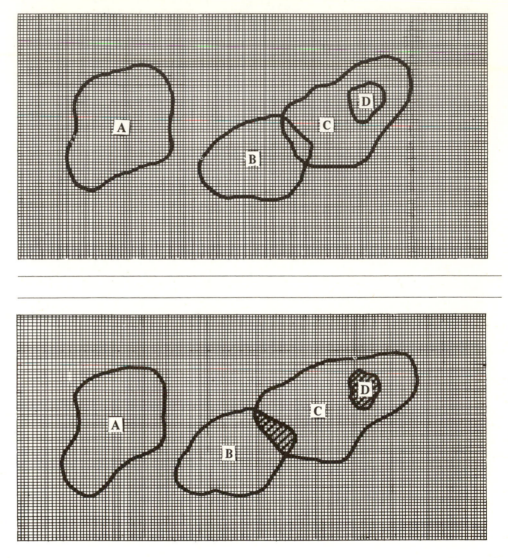

D is in danger of elimination

43 Clearly the greater the overlap in ecological niches, the greater the competition. Thus it is found that when any two species compete for the same niche, sooner or later one will oust the other, i.e. one niche – one species. Some organisms have a *broad niche specificity*. That is to say, they can use a wide variety of food materials (broad trophic niche) and can tolerate a wide range of environmental conditions (spatial niche). Others have a narrow niche specificity with very restricted diets and tolerating only a narrow band of environmental variation. Which of the organisms in frame 42 has the narrowest niche specificity?

44   Clearly if species D is to survive, it must entrench itself in its niche more efficiently than C. This will involve it in adjusting itself to a continually changing environment, e.g. daily and seasonal fluctuations in temperature, light, moisture, etc. The following extract from a paper by D. Koller* illustrates how organisms make this adjustment. How would you describe the mechanisms employed?

"Today, botanists are exploring a new realm of mechanisms that regulate the germination of plant seeds. These mechanisms help to determine the timing and locality of germination by restraining it in environments and seasons that do not afford a reasonable chance for the plant to complete a life-cycle 'from seed to seed'. Typical of these newly discovered mechanisms is that of chemical regulation, now under investigation in the Earhart Plant Research Laboratory of the California Institute of Technology and in the Department of Botany of the Hebrew University in Israel.

Clearly there is no worse place for a tomato or melon seed to germinate than inside the growing parent fruit; such vivipary would be highly disadvantageous. The warm, moist flesh of the fruit provides just the sort of environment in which the seeds might be expected to sprout, yet they rarely do. How is germination delayed until the fruit has fallen and decayed? The prevention of vivipary in most fleshy fruits is due to the presence in them of substances that specifically inhibit germination. Only when the seeds are free of the pulp and juice will they germinate.

More dramatic are the 'chemical rain-gauges' found in many dispersal units. These are inhibitory substances that are water-soluble, and are therefore readily leached out by rainfall. The amount of inhibitor in the dispersal unit is apparently adjusted so that the amount of rainfall needed to leach it out sufficiently to permit germination will at the same time moisten the soil sufficiently to ensure the plant's subsequent growth. In the dispersal units of wild smilograss (*Oryzopsis miliacea*), local varieties are 'gauged' to the rainfall pattern of their habitat. The importance of such rainfall-dependent germination control for the survival of plants in arid or semi-arid zones, where rainfall is limited and erratic, will be self-evident.

Another regulatory mechanism found in many dispersal units is the 'temperature gauge'. In its simplest form the temperature gauge restricts germination of a species to a specific temperature range that is often very narrow and precise. This then distinguishes plants that start their lives in cool climates and seasons from those that do so in warm ones. More highly developed regulation by temperature is found in plants that will germinate only when they are submitted to a specific change in temperature. Most common are the 'cold-requiring' seeds, the subject of extensive research at the Boyce Thompson Institute for Plant

*From 'Germination' by D. Koller. Copyright © 1959 by Scientific American. Inc. All rights reserved.

Research in Yonkers, N.Y. In order to germinate, these seeds require either one or two prolonged exposures (each of several weeks) to near-freezing temperatures, alternating with one or two exposures to higher temperatures. The apple, the peach and other plants that exhibit such mechanisms are invariably denizens of temperate climates; their ability to avoid germination before prolonged exposure to cold is of high survival value, since it minimizes the danger that seeds may germinate before the hazard has passed.

Moreover, like most other plants, temperate-zone plants are specific to one spatial niche, their entire development pattern (growth, flowering and fruiting) is synchronized with the climatic cycle to such an extent that they could not grow normally elsewhere. Their requirement for such 'seasonal thermoperiodicity' for germination is thus an important factor in assuring them a start in life in a suitable environment, that is one that includes a cold season. Unfortunately, we have no more than fragmentary knowledge of the physiological nature of this mechanism. But its complexity can be judged from the case of the snow trillium, the root growth of which is induced by a cold period, is carried forward in a following warm period and is not followed by shoot growth unless a second cold period intervenes.

Low temperature and light
. . . Common, though poorly understood . . . response to temperature variation is 'diurnal thermoperiodicity', a characteristic of plants that germinate far better under daily alternations of warmth and cold than they do at any constant temperature. Ecologically, such a mechanism can prevent germination in climates, seasons or soil depths where proper temperature alternations do not occur. Physiologically, we have almost no clue to the operation of the mechanism. Rhythmical (or cyclical) phenomena have been observed in many forms of living things: plants, mammals, birds, insects and micro-organisms. Many of these phenomena follow a 24-hour periodicity quite independent of the environmental, or astronomical, 24-hour cycle, but capable of being synchronized with it. It is quite likely that the study of this general phenomenon will lead to an understanding of diurnal thermo-periodicity in germination.

Dispersal units incorporate not only rainfall and temperature gauges, but also sensitive mechanisms that respond to light. Such a mechanism in the humble lettuce plant is the subject of research at three research institutions (the US Department of Agriculture laboratories at Beltsville, Md., the Hebrew University in Israel and the University of California at Los Angeles). In darkness lettuce seeds germinate tolerably well only within a narrow temperature range. Given light, they germinate promptly and uniformly over a very wide range, and under a variety of conditions that would absolutely inhibit germination in the dark. Dry lettuce seed is insensitive to light, but a few minutes after the seed is moistened it becomes light-sensitive, so sensitive that exposure for a few seconds to light with an intensity of a few foot-candles suffices to produce the full effect. The obvious analogy to photographic exposure extends further; if a soaked seed is exposed to light and then dried, it will retain the 'message' it received and, when it is subsequently remoistened, it will germinate in darkness.

A search of the light spectrum for the most effective wavelengths has shown that only the red portion of the visible spectrum stimulates germination. At the same time it was found that far-red light (on the boundary between the visible red and the infrared) is capable of reversing the stimulation by red light, thereby inhibiting germination. A flash of red stimulates germination. A flash of far-red, following closely, completely reverses the stimulation. This reversal is itself reversed when followed closely by another flash of red, and so on repeatedly. It is always the color of the final light-flash that is decisive. Our understanding of this mechanism is fragmentary. As in the case of the near-freezing of seeds, the results of the process are not immediately visible. We only perceive their end products, namely subsequent germination or non-germination in darkness.

Sensitivity to light implies the presence of a pigment that absorbs the light. The effects of the red and far-red indicate some properties of this pigment, but it has yet to be extracted, purified, identified and studied — a process that may take some time, since the pigment doubtless occurs in minute amounts. Luckily for the investigator, light-sensitive mechanisms of this kind are not restricted to seeds: an identical mechanism has been observed in many developmental processes of plants, and may also occur in animals. Thus etiolated plants (plants grown in darkness) will be pale, tall and spindly and will bear un-expanded leaves; upon exposure to light they begin to grow normally. Similarly the study of the relationship between flowering and the relative length of day and night has shown that, in order for the dark period to stimulate flowering in short-day plants or inhibit it in long-day plants, the darkness must not be interrupted by light. If the plant is exposed even briefly to low-intensity light near the middle of the dark period, the effect of the darkness on flowering may be wiped out. It turns out that in both etiolation and flowering the sensitivity to light responds to the same red and far red stimuli as germination.

It may be significant that gibberellin and another plant-growth substance, kinetin, simulate the red-light stimulus that triggers germination, but to complicate matters several substances (e.g. potassium nitrate and thiourea) which are not known as plant-growth regulators, also do so. Another complication is the fact that germination apparently loses its sensitivity to light when the embryos are decoated. It remains to be seen whether light acts on the embryo, somehow making it grow with vigor sufficient to overcome the resistance of the coat, or whether it works on an extra-embryonic entity, perhaps by activating an inhibitor in the coat.

Like a photographic plate, seeds can be over- and under-exposed. The brief flash of light that stimulates germination in lettuce and tobacco plants would be insufficient for the rush *Juncus maritimus*; on the other hand, although continuous illumination works as well as a flash in the case of lettuce, it would inhibit wild smilograss or the Negev saltbush, plants fully stimulated by a brief exposure. The finding that some seeds are as sensitive as fully mature plants to relative length of day and night does not, therefore, come as a surprise, in view of the fact that both have the same responses to the red and far red.

We can deduce some implications of this mechanism for the ecology of plants. Sensitivity to the period of light and dark may determine the season of germination just as it determines flower initiation and the on-set or end of dormancy in the buds of trees and shrubs. Inhibition by overexposure to light may be of value in preventing germination from occurring on an exposed solid surface, where treacherous conditions such as rapid drying or high temperature are common. This may be why the germination of many desert plants is inhibited by overexposure. Conversely, inhibition by underexposure may be of value in preventing germination from taking place in poorly illuminated or overpopulated localities. This may explain why many aquatic and marsh plants require light for germination."

Remember that the question asked was: What mechanisms do organisms possess which enable them to entrench themselves in their niches?

Change in the environment precipitates change in the concentration or structure of certain specific chemicals. The latter triggers off processes in the organism which adjust it to the new environmental conditions. It is a form of chemical feedback control.

45   In the case considered above we were thinking of plants. We saw how they adjusted to various physical changes in the environment. Animals have the additional problem of fluctuating food sources. Below is a graph showing the effect of variation in food supply on a primary consumer. Suggest, in the light of the previous frame, what mechanism may be operating here.

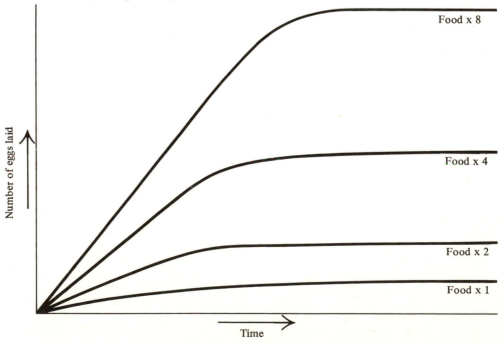

Feedback control. Higher food concentrations may trigger (chemically?) higher egg production; or lower food concentrations may trigger (chemically) the production of more and more inhibitor to egg production.

46 The following extract paper by Grummer & Beyer in *The Biology of Weeds*, ed. J.L. Harper (1960), suggests yet another mechanism by which an organism increases its competitive capability. Read the paper and answer the questions which follow at the end.

"In 1928 Prjanischikov and others demonstrated that the yield of flax is much reduced if a small percentage of *Camelina* is growing among the plants. This effect has been checked in our field experiments (Table 1). The effect of *Camelina* is much more pronounced than the effect of any other genus of weeds among flax. In pot-experiments twenty flax plants were combined with twenty plants of any weed. Whereas most weeds reduced the yield of flax from 20 to 30 per cent, the presence of *Camelina alyssum* resulted in a reduction of more than 80 per cent in the yield of flax (see Table 2). Different species of *Camelina* however gave nearly the same effect.

Table 1. *The reduction of flax-yield by* Camelina alyssum *and other weeds. Field-experiments 1956. Dry matter (kg).*

	Flax		Camelina		Total
	Seeds	Total	Seeds	Total	
Flax (control)	2.25 ±0.16	11.25 ± 0.50	– –	– –	11.52 ±0.51 –
Flax with natural weed flora	2.08 ±0.07	10.27 ± 0.07	– –	– –	10.27 ±0.08 –
Flax with 1% C. alyssum	2.25 ±0.05	13.45 ± 0.11	–	0.04 ±0.006	13.49 ±0.12 –
Flax with 5% C. alyssum	2.11 ±0.11	7.24 ± 0.14	0.07 –	0.46 ±0.01	7.70 ±0.13 –
Flax with 10% C. alyssum	1.52 ±0.05	6.03 ± 0.23	0.08 –	0.56 ±0.03	6.59 ±0.21 –

Table 2. *Reduction in yield of flax by various weeds*

	%
Flax without weeds	100.0
Flax with *Galium parisiense*	71.5
Flax with *Viola tricolor*	82.0
Flax with *Polygonum lapathifolium*	78.5
Flax with *Camelina alyssum*	16.4

At first, the production of poisonous substances by the roots was suspected. From the experiments carried out in vessels with nutrient solution no evidence could be found for the presence of toxic root excretions. Flax with *Camelina* in the same vessel grew as vigorously as flax alone. In the nutrient solution, conditions would have been favourable for the hypothetical exudates since the medium was kept sterile and no absorbents were present. We turned therefore to the hypothesis that leaves or other organs give off the poisonous substances. The influence of competition could easily be ruled out by specially designed experiments. [See table 3]. By a wall of plastic every pot was divided into two equal parts. Some of the pots contained *Camelina* and

Table 3.   *Yield of flax and* Camelina *in pot experiments exposed out of doors*

Design of experiment	Weight of seeds produced (g)	Total dry weight per plant (g)
20 plants of flax and 20 plants of *Camelina* arranged in a chessboard pattern		
Flax	0.62	2.35
Camelina	1.33	4.63
Total	1.95	6.98
Controls. Pure stands of flax:		
20 flax plants per pot	2.73	10.68
40 flax plants per pot	2.47	12.63

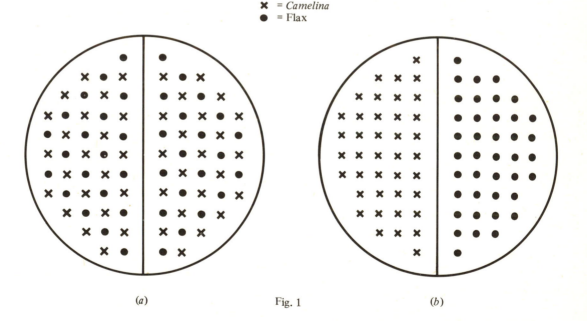

✖ = *Camelina*
● = Flax

(a)   Fig. 1   (b)

flax on both sides of the wall (Fig. 1*a*) so that root competition was possible. Other pots contained *Camelina* on one side and flax on the other side of the wall (Fig. 1*b*). No root competition was possible.

Moreover, some of the pots were kept in the open air and were subject to natural rain. Others were kept in the greenhouse, some of them with artificial irrigation. In others, only the soil was moistened but all leaves were carefully protected from rain or irrigation.

With natural or artificial rain and no root competition the yield of flax was reduced. Additional competition resulted in a further reduction. When no rain fell on the leaves the competition was entirely ruled out, flax grew as well as in control pots.

Competition could also be eliminated by experiments in test tubes. Tubes were filled with soil and one plant was placed in each tube. Every plant had an independent nutrient supply, and no competition for water or nutrients was possible. With artificial rain the flax plants in the immediate neighbourhood of *Camelina* plants produced 40 per cent less dry matter than the controls in which no rain was allowed to fall on the leaves, but instead the same amount of water was applied directly to the soil (Table 4)."

Table 4. *Reduction in the yield of flax due to* Camelina. *Greenhouse experiments. Every plant in a separate test tube of soil. No competition for water and nutrients.*

	Mean dry weight (g)	
	Flax	*Camelina*
17 test tubes with flax, 17 with *Camelina*, artificial rain on the soil, leaves kept dry	(a) 540 ± 14.8 (b) 486 ± 14.2 (c) 564 ± 15.4	492 ± 13.1 437 ± 12.9 468 ± 12.6
Mean	530 ± 7.8	466 ± 7.1
17 test tubes with flax, 17 with *Camelina*, artificial rain on leaves	(a) 337 ± 11.2 (b) 341 ± 10.6 (c) 283 ± 13.4	455 ± 10.3 435 ± 10.1 492 ± 13.1
Mean	320 ± 6.4	461 ± 5.9

Questions
1. What theory is suggested by these results about the way in which *Camelina* inhibits the yield of flax?
2. By how much does the natural weed flora reduce the yield of flax?
3. By how much does *Camelina* reduce the yield of flax?
4. How was it possible experimentally to decide whether root competition for nutrients and water was occurring?

---

1. The results of these investigations strongly support the theory that toxic substances are washed out of the *Camelina* leaves by rain and that these substances play a role in the reduction of flax yielded.
2. Between 20 and 30 per cent.
3. Over 80 per cent.
4. (a) By comparing the results obtained from a 'chessboard' arrangement of *Camelina* and flax with a control in which the soil in the pot was divided into two halves by a plastic sheet, so that the roots of *Camelina* were clearly separated from those of flax.
   (b) By growing *Camelina* and flax in a separate test tube each

with their own water and nutrient supply.

In both (*a*) and (*b*) comparison of the effects of natural rainwater and artificial irrigation could be made on the growth of flax in close proximity to *Camelina*.

---

47 Let us now consider an extract from another paper on this theme. It is by G. Deleuil (1954; *Comptes Rendus de l'Académie des Sciences*). What different mechanisms are at work in this case?

"Following my researches on toxic secretions by roots (root exudates), my attention was drawn to the curious distribution of three species of plants present on the south coast of the Masil de la Nerthe to the west of Marseilles. The species in question were *Allium chamaemoly*, *Hyoseris scabra* and *Bellis annua*.

We noted the following distributions:

1. The three species were found grouped together in a circle of radius 20 cm.
2. The following species were found grouped in pairs: *Allium chamaemoly* with *Bellis annua* and *Bellis annua* with *Hyoseris scabra*.
3. *Allium chamaemoly* and *Hyoseris scabra* were *never* found together on their own.

I then conducted the following experiments to see if these peculiar distributions could be explained in terms of toxic root exudates. Sterilized soil was used in all the experiments.

1. Seeds of *Hyoseris scabra* were sown in pots containing *Allium chamaemoly* plants. The seeds germinated normally, but the seedlings withered and died. The same results were obtained if, instead of sowing the seeds in the presence of *Allium chamaemoly*, one watered a pure sowing of *Hyoseris scabra* with water draining from *Allium chamaemoly* roots (i.e. *A.c* leachate). A pure sowing of *Hyoseris scabra* provided with ordinary water acted as a control and developed normally.
2. *Bellis annua* seeds were sown in pots containing *Allium chamaemoly* plants. The seeds germinated and, after a struggle, the plants established themselves. As soon as the young plants were well developed, *Hyoseris scabra* seeds were sown among them. This time the *Hyoseris scabra* seedlings developed normally, and the three species lived together without harming one another.
3. *Hyoseris scabra* seeds were sown in a pot containing *Allium chamaemoly* plants. When all the seeds were watered with *B.a* leachate, different phenomena were observed depending on whether the *Bellis annua* plants had been collected from the immediate vicinity of *Allium chamaemoly* or from where *Allium chamaemoly* was absent. In the latter case the leachate had no effect, the *Hyoseris scabra* seedlings dying as in experiment one. On the other hand, when the *Bellis annua* plants had been grown in the presence of *Allium chamaemoly*, its leachate facilitated complete development of the *Hyoseris scabra* seedlings.
4. *Bellis annua* seedlings were sown in a pot containing soil in which *Allium chamaemoly* had been grown, but removed at maturity.

The leachate from these *Bellis annua* plants possessed the same properties with regard to *Hyoseris scabra* seedlings as those possessed by *Bellis annua* plants grown in the actual presence of *Allium chamaemoly* plants."

Hint. In answering the question posed above (i.e. What different mechanisms are at work?), consider first the implications of each experiment before coming to a general conclusion.

From Experiment 1. *Allium chamaemoly* secretes a toxic substance which destroys *Hyoseris scabra* seedlings.

From Experiment 2. *Hyoseris scabra* can develop and live normally in the presence of *Allium chamaemoly* providing *Bellis annua* is present. *Bellis annua* can germinate and develop in the presence of *Allium chamaemoly*.

From Experiment 3. *Bellis annua* 'detects' the presence of *Allium chamaemoly* toxins, and produces a neutralizing substance (antitoxin) in response. *Hyoseris scabra* plants can survive in this environment.

From Experiment 4. *Bellis annua* antitoxin is produced in response to the presence of toxins emitted into the soil by *Allium chamaemoly*.

General conclusion. *Allium chamaemoly* increases its competitive capability by modifying its environment. *Bellis annua*, by virtue of a feedback control mechanism, establishes itself within what is potentially the *Allium chamaemoly* niche. *Hyoseris scabra* also establishes itself, under the protection of both the other species.

---

48   Competitive situations are also known in the animal kingdom. The following short account illustrates how ants not only tolerate certain 'intruding' beetle species, but also feed and shelter them. For a full account you are recommended to read *Communication between ants and their guests* by B. Höllander (*Scientific American* **224**, 1971, pp. 86–93).

Most ants are social insects, and as such possess a sophisticated internal communication system which enables a colony to carry out its collaborative activities in nest-building, food-gathering, care of young and defence against enemies.

A number of ants are remarkable 'farmers'. Some species for example, grow fungal crops for food within certain areas of their nests; many other species are well known for their herding, protection and 'milking' of aphids (greenfly and blackfly), on plants outside, but near to their nests. The attraction here for the ants is the sugary secretion called honeydew, which the aphids produce as a result of their rich carbohydrate diet of plant sap. Such behaviour seems remarkably clever and sensible by human standards. What is difficult to understand, however, is the thriving parasitic relationship which not only exists between certain species of beetles and ants, but is almost cordially welcomed by the ants. How is it that, given the social structure of the ant colony, the intruding insect not only makes its home in the ants nest (actually in the brood chamber) but then coolly eats the host ants' young?

One example will illustrate the probable answer to this question. A European beetle *Atemeles pubicollis*, spends the larval stage of its life cycle in the nest of the mound-making ant *Formica plyctena*. The adoption of the beetle in the first place depends on chemical communication. The larva secretes a chemical substance externally from glands in its integument since the substance imitates a pheromone (ectohormone, see Unit 4, Book 11) that ant larvae themselves produce to release brooding behaviour in adult ants, the beetle larvae are attractive and acceptable to the adult ants. Once 'adopted' by the ants they are then taken into the brood chamber of the nest.

To elicit a larval feeding response from the ant, the beetle larvae use a different form of communication. This time the beetle larvae imitate the begging behaviour of the ant larvae, but perform the begging even more intensely than the ant larvae and in consequence receive more food — deceit indeed!!

In addition to the food which they receive from the adult ants, the beetle larvae are voracious predators on the ant larvae in the brood chamber. This being the case, one may well ask how the ant colony manages to survive.

The answer seems to be that the beetle larvae are cannibalistic and cannot distinguish by odour fellow beetle larvae from ant larvae. As a result beetle larvae cut down their own numbers, whereas the ant larvae do not.

Extraordinary as it may seem, the *Atemeles* beetles have two homes with ants — a summer 'residence' and a winter 'residence', when the beetle larvae have pupated and hatched in the *Formica* (ant) nest, the adult beetles migrate in the autumn to nests of the insect-eating ants of the genus *Myrmica*. Since these ants live in open grassland, migration of the beetles from woodland to grassland is thought to be brought about by a response towards increasing light intensity.

Before it leaves the *Formica* nest, the *Atemeles* beetle adds insult to injury by having the audacity to beg for a food supply for its autumnal migration from its host. Here tactile stimulation is involved. The beetle knocks rapidly on an ant with its antennae to attract attention, and then induces the ant to regurgitate food by touching the ant's mouthparts with its maxillae (mouthparts) and forelegs.

The reason for the migration (fiendish things these beetles!) is that brood-keeping and food supply are maintained in *Myrmica* colonies throughout the winter, whereas *Formica* suspend raising their young.

In the *Myrmica* nests, the beetles still sexually immature, can be fed and ripen to maturity by the spring, at which time they return to the *Formica* nests for mating and laying their eggs. This remarkably advanced form of adaptation by *Atemeles* beetles allows them to take maximum advantage of the social life of each of the two ant species that serve as hosts.

We see, therefore, that various behaviour patterns release a defensive response on the part of the occupant of the niche (i.e. the ant). How many different forms of communication (behaviour) patterns were used by the *Atemeles* beetle to establish its position within the same niche as the ant?

___

(*a*)   Chemical communication (adoption of the larva)
(*b*)   'Sign' communication — i.e. copying the begging behaviour of ant larvae (but more so) to obtain food.
(*c*)   Tactile communication or stimulation of the adult to regurgitate food in preparation for the autumnal migration.

___

49   Let us now return to more general considerations of ecosystems. With increasing maturity of an ecosystem increasing quantities of the nutrients are lodged in a more or less static biomass. Species diversity also increases with maturity. At which stage of a succession (mature or immature) would you expect success to come to organisms exhibiting
(*a*)   rapid growth reproduction;
(*b*)   Growth and reproduction subject to feedback control?
Why?

___

(*a*)   Immature ecosystem. Since there will be relatively few species present, success will depend on the monopoly of available resources.
(*b*)   Mature ecosystem. With numerous species present success in competition will depend on precise adaptation to specific conditions.

___

50 This favouring of particular species characteristics by particular environments is called '*selection pressure*'. Thus there is a strong selection pressure exerted by the environment obtaining in immature ecosystems which favours organisms with rapid growth and reproduction, whereas, in a mature ecosystem, selection pressure favours organisms whose growth and reproduction are regulated by feedback controls. Therefore in the first case 'quantity' production is selected for and in the second case, 'quality' production.
At which stage of ecosystem development (immature or mature) would you expect life cycles (birth to reproduction) to be short and simple? Why?

Immature, since a rapid build up of biomass has priority. Organisms in mature ecosystems are often characterized by long, complex life cycles, e.g. the moth which has a feeding stage (larva), a resting stage (pupa), and a breeding stage (imago or adult moth).

51 In the light of the last two frames, at what successional stages would you expect successful organisms to possess:
(*a*) narrow niche specificities;
(*b*) broad niche specificities?

(*a*) Mature stages
(*b*) Immature stages

52 The meaning of the term symbiosis (common life) is sometimes broadened to include all kinds and degrees of partnership between organisms of different species, e.g. commensalism, 'strict' symbiosis, predation, parasitism, etc. Would you expect symbiosis to be *more* or *less* characteristic of the ecosystem with increasing maturity?

More characteristic, since conditions which favour species with narrow niche specificities, are also likely to favour symbiotic relationships.

53 Clearly the more intimate the relationship between organisms the more important do feedback controls become. Would you therefore expect biochemical diversity (the range of chemicals present in the ecosystem) to *increase* or *decrease* with maturity?

Increase. This has been found to be the case in laboratory ecosystems.

54 Loss of the essential nutrient chemicals from the ecosystem also represent a potential hazard to the stability of the ecosystem. It is found, however, that the communities of mature ecosystems have a very slow exchange rate with their physical environment as compared with developing ecosystems. Furthermore, in the mature ecosystem as we have seen, the nutrients are located largely in the community biomass. The net result of these properties of mature systems is that loss of nutrients from the ecosystem is far less than in immature systems, i.e. the nutrient cycles are *closed* in the former and *open* in the latter. This is illustrated below.

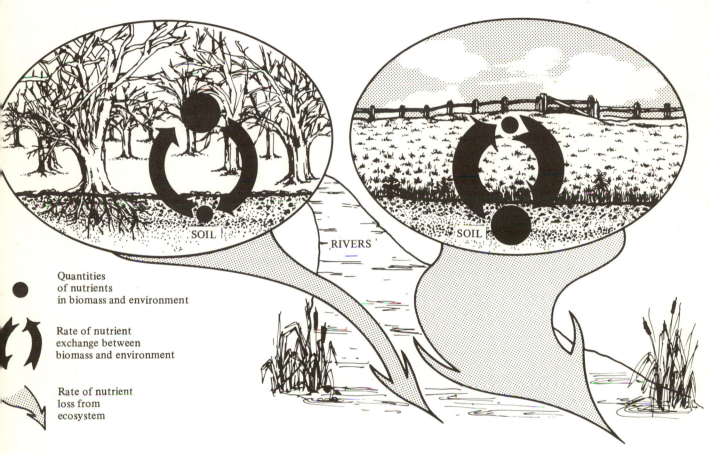

MATURE FOREST ECOSYSTEM

IMMATURE GRASSLAND ECOSYSTEM

SOIL

SOIL

RIVERS

● Quantities of nutrients in biomass and environment

↻ Rate of nutrient exchange between biomass and environment

⇘ Rate of nutrient loss from ecosystem

What properties does the mature ecosystem possess which may be responsible for its capacity for slow nutrient cycling and small nutrient loss?

The larger size and longevity of the organisms permits safe storage of nutrients. This, coupled with sophisticated feedback mechanisms, enables the ecosystem to regulate the mobilization and movements of stored materials very precisely.

55 We may now summarize our observations on change in an ecosystem.

	Stages of ecosystem development	
	Immature	Mature
Community energetics		
P/R	> 1	Approaches 1
Net production	High	Low
Food chains	Linear, grazing	Web-like, detritus-feeding
Community structure		
Total organic matter	Small	Large
Inorganic nutrients	Extrabiotic	Intrabiotic
Species diversity	Low	High
Biochemical diversity	Low	High
Stratification	Poorly organized	Well organized
Life history		
Niche specialization	Broad	Narrow
Size of organism	Small	Large
Life cycles	Short, simple	Long, complex
Nutrient cycling		
Mineral cycles	Open	Closed
Nutrient exchange rate between organisms and environment	Rapid	Slow
Role of detritus in nutrient regeneration	Unimportant	Important
Selection pressure		
Growth form	For rapid growth	For feedback control
Production	Quantity	Quality

Now, in conclusion, we come to the most important question.
The overall trend in the succession of an ecosystem is the formation of as large and as diverse a structure as the energy input and physical constraints will permit. The mature ecosystem has been shown to be remarkably stable.
Which of all the characteristics we have observed are responsible for this stability?

The existence of feedback controls, which eliminate the effects of physical and biotic oscillations and which indirectly facilitate nutrient conservation within the ecosystem

In the next section, we shall consider some of the ways in which man both deliberately and unconsciously involves himself in the fate of ecosystems and raise questions about the wisdom and propriety of his practice.

4.4. Man and ecosystems

4.4.1. Overview

This section is not programmed. Instead it consists of selected short commentaries or reviews by eminent people in the field of ecology.*

The first section looks at ecosystems in jeopardy, coupled with the problems of recycling our resources, pollution, and increase in world population.

The last two papers are included to show what can be done to improve the quality of the environment, if we put our minds to it.

4.4.2. Objectives

At the end of this section you should be able to:
1. Describe giving examples, some consequences of the mismanagement of ecosystems by man.
2. Explain the implications of an exponentially increasing human population on the demand for food, space and resources.
3. Describe two methods by which scientists are trying to improve the quality of some parts of our polluted environment.

4.4.3. Ecosystems in jeopardy †

The plants, animals, and microorganisms that live in an area and make up a biological community are interconnected by an intricate web of relationships, which includes the physical environment in which these organisms exist. These interdependent biological and physical components make up what biologists call an ecosystem. The ecosystem concept emphasizes the functional relationships among organisms and between organisms and their physical environments. These functional relationships are exemplified by the food chains through which energy flows in ecosystems, as well as by the pathways along which the chemical elements essential to life move through the ecosystem. These pathways are generally circular; the elements pass through the system in cycles. The cycling of some elements is so slow, however, that in the time span of interest to us, movement appears to be one-way. An understanding of the flow of energy and the cycling of materials in ecosystems is essential to our perception of what is perhaps the most subtle and dangerous threat to man's existence. This threat is the potential destruction by man's own activities, of those ecological systems upon which the very existence of the human species depends.

Food webs
The food web of a Long Island estuary has been thoroughly investigated by biologists George M. Woodwell, Charles F. Wurster, and Peter A. Isaacson. The relationships they discovered are illustrated in Figure [1]; their study illustrates several important characteristics of most food webs. One is complexity. Although only some of the kinds of plants and animals in this ecosystem are shown in this figure, it is evident that most of the consumers

*We are extremely grateful to the authors and publishers for giving us permission to quote their work.

†Extracts from *Population, Resources, Environment: Issues in Human Ecology*, 2nd edition, Paul R. Ehrlich & Anne H. Ehrlich. W.H. Freeman & Co. Copyright © 1972.

112

chain would have disastrous consequences for the entire ecosystem. Suppose, attacked by more than one predator. To put it another way, the food chains are interlinked. Ecologists believe that complexity is in part responsible for the stability of most ecosystems. Apparently, the more food chains there are in an ecosystem and the more cross-connecting links there are among them, the more chances there are for the ecosystem to compensate for changes imposed upon it.

For example, suppose that the marsh plant–cricket–redwing blackbird section of Figure 1 represented an isolated entire ecosystem. If that were the case, removing the blackbirds – say, by shooting – would lead to a cricket plague. This in turn might lead to the defoliation of the plants, and then to the starvation of the crickets. In short, a change in one link of such a simple

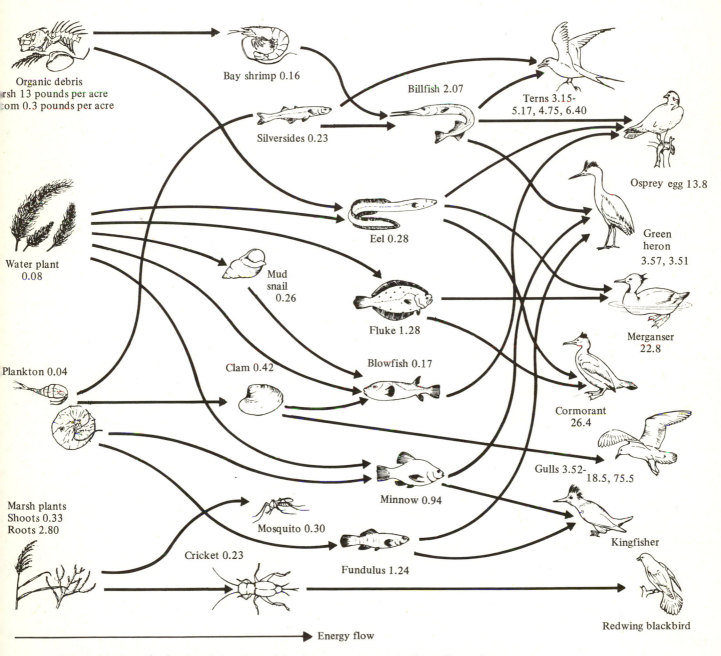

Fig. 1. Portion of a food web in a Long Island estuary. Arrows indicate flow of energy. Numbers are the parts per million of DDT found in each kind of organism. (From Woodwell, *Toxic Substances and Ecological Cycles*. Copyright © 1967 by Scientific American, Inc. All rights reserved.)

feed on several different organisms, and that most prey organisms are however, that the cormorants were removed from the larger system. Populations of flukes and eels would probably increase, which in turn might reduce the population of green algae (*Cladophora*). But there would be more food for mergansers and ospreys, and their populations would probably enlarge, leading to a reduction of eels and flukes. In turn, the algae would recover.

Needless to say, things do not normally happen that simply and neatly in nature. But we have both observational and theoretical reasons to believe that the general principle holds: complexity is an important factor in producing stability. Complex communities, such as the deciduous forests that cover much of the eastern United States, persist year after year if man does not interfere with them. An oak–hickory forest is quite stable in comparison with an ultrasimplified community, such as a cornfield, which is a man-made stand of a single kind of grass. A cornfield has little natural stability and is subject to almost instant ruin if it is not constantly managed by man. Similarly, arctic and subarctic ecosystems, which are characterized by simplicity, tend to be less stable than complex tropical forest ecosystems. In arctic regions the instability is manifested in frequent, violent fluctuations in the populations of such northern animals as lemmings, hares, and foxes. In contrast, outbreaks of one species do not occur as often in complex tropical forests. Ecologist Robert MacArthur suggested in 1955 that the stability of an ecosystem is a function of the number of links in the web of food chains. He developed a measure of that stability using information theory. Although the bases of stability now appear somewhat more complicated than those proposed by MacArthur in his pioneering work, the idea that complexity promotes stability still appears to be theoretically sound.

Concentration of toxic substances in ecosystems
Nowhere is man's ecological naivete more evident than in his assumptions about the capacity of the atmosphere, soils, rivers, and oceans to absorb pollution. These assumptions all too often take the following form: if one gallon of poison is added to one billion gallons of water, then the highest concentration of poison to which anything will be exposed is about one part per billion. This might be approximately true if complete mixing by diffusion took place rapidly, which it often does not, and *if only physical systems were involved*. But because *biological systems* are involved, the situation is radically different. For example, filter-feeding animals may concentrate poisons to levels far higher than those found in the surrounding medium. Oysters live in shallow water near the shore, where pollution is heaviest. Consequently, their bodies often contain much higher concentrations of radioactive substances or lethal chemicals than the water in which they live. For instance, they have been found to accumulate up to 70 000 times the concentration of chlorinated hydrocarbon insecticides found in their environment. Food chains may lead to the concentrations of toxic substances; as plant physiologist Barry Commoner of Washington University once put it, they act as a kind of 'biological amplifier'. The diagram of the Long Island estuary food web (Fig. 1) shows how the concentration of DDT and its derivatives tend to increase in food chains from one trophic level to another. This tendency is especially marked for the chlorinated hydrocarbons because of their high solubility in fatty substances and their low water solubility. Although the clam and the mud snail are at the same trophic level, the filter-feeding clam accumulates more than half again as much DDT as the

mud snail because of the differences in their food-capturing habits.

The mechanism of concentration is simple. Because of the Second Law of Thermodynamics, the mass of herbivores normally cannot be as great as the mass of plants they feed on. With each step upward in a food chain the *biomass* is reduced. Energy present in the chemical bonds of organisms at one level does not all end up as bond energy at the next level, because much of the energy is degraded to heat at each step. In contrast, losses of DDT and related compounds along a food chain are small compared to the amount that is transferred upward through the chain. As a result, the concentration of DDT increases at each level. Concentrations in the birds at the end of the food chain are from tens to many hundreds of times as high as they are in the animals farther down in the chain. In predatory birds, the concentration of DDT may be a *million* times as high as that in estuarine waters.

Clear Lake, in California, has long been a favorite of fishermen, and now attracts water-skiers and vacationers of all kinds. Unfortunately, a midge (known locally as the Clear Lake 'gnat') reproduces in great numbers in certain years. The insect is considered a pest merely because it is phototropic (attracted to light), and for no other reason. In an attempt to control the gnat, a program was begun in the late summer of 1949 using DDD, a less toxic but equally persistent relative of DDT. The rate of application to the lake water was 14 parts per billion (p.p.b.). The first application of what was then thought to be a relatively harmless pesticide eliminated about 99 per cent of the gnats, as did the next application of 20 p.p.b. in 1954. By the time the lake was treated for the third and last time in 1957, the gnat plus many other species of insects and other pests had developed some resistance to the pesticide. (It should be pointed out that within two weeks after each treatment, no DDD could be detected in the lake waters.)

Before 1950 Clear Lake had been a nesting ground for about 1000 pairs of western grebes (duck-like diving birds that feed primarily on small fishes and other aquatic organisms). Not only did many grebes die soon after the 1954 and 1957 treatments, but fairly large die-offs occurred in subsequent years. Furthermore, the survivors were unable to reproduce. From 1950 to 1961 no young were produced; in 1962 a single grebe hatched. Reproduction remained unsuccessful until 1969, many years after the first introduction of DDD into the lake. Studies designed to determine the concentration of DDD from the lowest trophic levels to the highest revealed that the microscopic plankton of the lake contained about 250 times that of application (the original concentration in the lake water). The concentration in frogs was 2000 times that of application, in sunfish, about 12 000 and in grebes, as high as 80 000 times. The figures given for frogs and the other animals higher in the food chain are for the visceral fat. The flesh of several species of game fish were also examined. Some white catfish contained almost 10 000 times the original concentration of DDD in the water. These data make it obvious why no DDD could be detected in the lake water only two weeks after application: because of its high solubility in the fatty materials of biological systems, the insecticide had been absorbed almost completely by the *living* components of the lake's ecosystem.

Nitrogen, phosphates, and ecosystems

Much of the nitrogen in natural soils is contained in humus, the organic matter of the soil. Humus is a poorly understood complex of compounds of high molecular weight. Inorganic nitrogen in such soils normally accounts for less than 2 per cent of the nitrogen present; often the majority is tied up in the large organic molecules of humus, which are derived from such varied sources as the fibrous remains of woody plant tissues, insect skeletons, and animal manure. These substances, in addition to their chemical value, increase the capacity of the soil to retain water. The presence of humus makes the soil a favourable medium for the complicated chemical reactions and mineral transport needed for the growth of higher plants. Bacteria in the soil decompose humus to form nitrates and other nutrient substances required by plant roots.

Roots require oxygen in order to do the work necessary for the uptake of nitrates and other nutrients, but oxygen is not available if the soil is tightly compacted. Thus another important benefit of humus is to maintain soil porosity and so permit oxygen to penetrate to the roots of plants.

In natural soil systems the nitrogen cycle is 'tight'. Not much nitrogen is removed from the soil by leaching or surface run-off. It has been shown experimentally that by maintaining the supply of humus the fertility of the soil can be perpetuated. This is not possible when fertilizers containing inorganic nitrogen are employed, unless organic carbon (in such forms as sawdust or straw) is supplied to the soil microorganisms. The undesirable decline of humus which often occurs under inorganic fertilization is due to the failure of the farmer to return crop residues (and thus carbon) to his fields. The decline is not caused by any deficiency in the fertilizers themselves. Indeed, if carbon is supplied in the proper proportion with inorganic nitrogen, the supply of humus can be increased and the quality of the soil improved.

If attempts are made to maintain soil fertility by continued applications of inorganic nitrogen fertilizers alone, the capacity of the soil to retain nitrogen is reduced as its humus content drops. In humus, nitrogen is combined into nonsoluble forms that are not leached from the soil by rainwater. Depletion of humus 'loosens' the soil cycles and permits large amounts of nitrate to be flushed into rivers and lakes. The use of inorganic fertilizers in the United States has been multiplied some 12-fold in the past 25 years. One result of this dramatic increase has been a concomitant rise in the content of nitrate in surface water, atmosphere, and rain. Another has been a 50 per cent reduction of the original organic nitrogen content of Midwestern soils.

The results of the added nitrogen content of our waters are exemplified in part by the now well-documented fate of Lake Erie. The waters of Lake Erie are so polluted that the US Public Health Service has urged ships on the lake not to use lake water taken within 5 miles of the United States shore. The water is so badly contaminated that neither boiling nor chlorination will purify it; although the organisms in it would be killed, the dangerous chemicals it contains would not be removed or broken down.

The sources of Lake Erie's pollution are many. A report to the Federal Water Pollution Control Agency cites as the main source of pollution the raw sewage dumped into the lake by lakeside municipalities, especially Cleveland, Toledo, and Euclid, Ohio; and Wayne County (Detroit), Michigan. The report cites industry as another major source of pollution, and names the Ford Motor Company, Republic Steel, and Bethlehem Steel as significant pollutors. Finally, the basin of Lake Erie contains an estimated 30 000 square miles of farmland, and another important source of pollution is run-off from these farmlands.

Let us examine the last source first. The waters draining the farmlands of the Middle West are rich in nitrogen as a result of the heavy use of inorganic nitrogen fertilizers. Indeed, they have an estimated nitrogen content equivalent to the sewage of some 20 000 000 people — *about twice the total human population of the Lake Erie basin*. Thus, besides fertilizing their farms with nitrogen, the farmers are also fertilizing Lake Erie; their nitrogen contribution is of the same order of magnitude as that of the municipalities and the industrial pollutors. The nitrogen balance of the lake has been seriously disturbed, and the abundance of inorganic nitrates encourages the growth of certain algae. In recent years these algae have produced monstrous blooms — big masses of algae that grow extremely quickly, cover huge areas, foul beaches, and then die.

The bacterial decay of these masses of algae consumes oxygen, reducing the amount of oxygen available for fishes and other animals. Such blooms and oxygen depletions are characteristic of lakes undergoing *eutrophication*, which may be loosely translated as 'overfertilization'. Phosphate levels in US surface waters have increased 27-fold in recent years. Phosphates, as well as nitrates, are implicated in this problem. Sources include fertilizer run-off and industrial waste, but 60 per cent of the phosphate entering US waters comes from municipal sewage. The primary source of phosphates in sewage is household detergents.

The basic sequence of eutrophication is simple in outline. Inorganic nitrates and phosphates are washed into the lake. These inorganic chemicals are converted into organic forms as huge blooms of algae develop; the subsequent decomposition of the algae depletes water of oxygen and kills off animals that have high oxygen requirements. Much of the nitrate and phosphate remains in the lake, settling to the bottom with the decaying mass of algae. The bottom of Lake Erie now has a layer of muck that varies from 20–125 feet in thickness; this layer is immensely rich in phosphorus and nitrogen compounds. These compounds are bound by a 'skin' of insoluble iron compounds that covers the mud. Unfortunately, the iron compounds change to a more soluble form in the absence of oxygen. Thus the oxygen depletion itself may cause the release into the lake of more of the nutrients responsible for the lake's troubles, and eutrophication may take place even more rapidly. Barry Commoner believes that if this breaking down of the mud skin should continue and result in the release of large amounts of nitrogen and phosphorus, the lake may face a disaster that would dwarf its present troubles.

Lake Erie is just one outstanding example of a general problem that is well known to most Americans — the gross pollution of our lakes, rivers and streams. All manner of organic and inorganic wastes end up in our inland waters: raw sewage, manure, paunch manure (the stomach contents of slaughtered animals), detergents, acids, pesticides, garbage; the list goes on and on. All of these substances affect the life in the water, all too often exterminating much of it, and at the very least modifying the ecosystems in profound ways. This pollution problem is worldwide; many of the rivers of the Earth are quickly approaching the 'too thin to plow and too thick to drink' stage.

In the United States and some other areas, serious attempts have been made to clean up fresh-water systems. These have met with mixed success. It is impossible to know whether we are gaining, holding our own, or losing at the moment, but the situation is bad, and the outlook for the future in the United States is not encouraging. Even an isolated beauty like Lake Tahoe, a high Sierra lake shared by California and Nevada, is threatened. Barry Commoner has estimated that by 1980 urban sewage alone, if left untreated, would consume all the available oxygen in all 25 major river systems of the nation. The eutrophication problems in the US have now even spread to estuarine and inshore waters where sewage is dumped. Many oyster and clam beds have been damaged or destroyed, and some fisheries have all but disappeared. Oxygen depletion and accompanying changes in water quality have been shown to induce marked destabilizing changes in local marine ecosystems.

Similar problems exist in many other parts of the world. Lake Baikal in the Soviet Union seems slowly to be heading for a fate similar to Lake Erie's, despite the protests of Russian conservationists. Many lakes and rivers in Europe and Asia are beginning to show signs of eutrophication, often within 10–20 years after the start of human pollution. The rivers in Italy are so badly polluted that Italian scientists fear that marine life in the Mediterranean is endangered. In most UDCs, rivers are simply open sewers. Eutrophication is not usually a problem there, except in a few areas where some industrial development has taken place. However, the Green Revolution, with its required high levels of fertilization, may change the situation.

Inorganic nitrate and phosphate fertilizers must be considered a technological success because they do succeed in raising the amounts of free nutrients in the soil, but it is precisely this success that has led to eutrophication as those nutrients are leached out of the soil by ground water. It has been predicted that in 25 to 50 years the ultimate crisis in agriculture will occur in the United States. Either the fertility of the soil will drop precipitously, because inorganic fertilizers will be withheld, throwing the nation into a food crisis, or the amounts of inorganic nitrates and phosphates applied to the land will be so large as to cause an insoluble water pollution problem. One would hope that before this comes to pass the rate of fertilizer use will be moderated and laws will require return of plant residues or other means of supplying the organic carbon necessary to build humus. Even these steps might not avert a water crisis produced by two other technological successes: high-compression automobile engines, which produce an inorganic nitrogen fallout, and modern sewage treatment plants, which produce an effluent rich in inorganic nitrates and phosphates. The significance of these inorganic nutrients has only recently been widely recognized.

In the light of these and many other assaults on the environment, it would behove us to begin immediately to head off future threats. As is becoming apparent, use of the internal combustion engine will have to be greatly reduced. It is imperative that either our present sewage plants be completely redesigned to eliminate nutrients from the effluent, or that a way be found to reclaim the nutrients for fertilizer. Also needed are new sewage plants for the many communities that still pour raw sewage into our waters. Such a program requires money and effort, some of which might be provided immediately by employing the Army Corps of Engineers and the Bureau of Reclamation for such projects.

The problem of controlling the run-off of nutrients from farms is more difficult, but it is known that nitrates leach from soil because they are anions (negatively charged groups of atoms), and the capacity of the soil to retain anions is low. Mixing a resin with high binding affinity for anions into the soil would increase its capacity to hold nitrates. Certainly, experimental work in this area should be initiated immediately, but 'solutions' of this sort must be monitored very carefully. Often they have a tendency to create problems more serious than those they solve.

A more immediately available approach to controlling agricultural run-off might be to halt by law the practice of handling manure from farm animals as a waste product. Roughly 80 per cent of American cattle are produced on feedlots, and most of their manure is treated as sewage, which more than doubles the sewage volume of the nation. Even if costs prove to be higher, manure should be returned to the land to help build humus.

Sterile, concentrated sewage (sludge) can also be used to improve soil. Sludge from Chicago is now being employed to restore stripmined land in southern Illinois. Crops are being grown very successfully on the reclaimed land. Recycled sewage has long been used as fertilizer in England, Australia, and other countries with excellent results. The prejudices of American farmers are the major obstacle to the development of similar practices in the United States.

Now test your understanding of the subject matter in this article by answering the following questions.

1   What factors are likely to give rise to the considerable variation in the amounts of DDT (p.p.m.) found in the tertiary consumers of a food web?

2   What is eutrophication? Why is it a problem? Explain how the use of humus and organic manure can maintain soil fertility better than continuous applications of inorganic fertilizers alone.

4.4.4. Mineral resources*

Until recently little attention was paid to conservation of mineral resources because it was assumed that there were plenty for centuries to come and that nothing could be done to save them anyway. *It is now apparent that both assumptions are dead wrong!* Cloud (1968, 1969, 1970) has inventoried supplies and reviewed prospects. He introduces two concepts (in his 1969 paper) that are useful in evaluating the situation. The first is the *demographic quotient*, which we shall designate as 'Q':

$$Q = \frac{\text{total resources available}}{\text{population density} \times \text{per capita consumption}}$$

As this quotient goes down, so does the quality of modern life; it is going down at a frightening rate because available supplies can only (or eventually) go down as consumption goes up. Even if available resources could be kept constant by recycle or other means, the situation deteriorates as long as population, and especially per capita consumption, increases at a rapid rate. Thus, in the United States economic and technological growth based on exploitation of natural resources is increasing at a rate of 10 per cent per year (doubling time is about 7 years!), urban growth is increasing 6 per cent per year, while the population growth is only about 1 per cent. If the undeveloped world with its huge populations were to increase its per capita use of minerals (and the fossil fuels required to extract and use the mineral

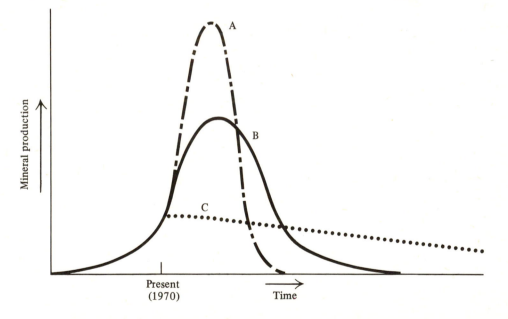

Fig. 1. Alternate depletion patterns for mineral resources. A. Pattern of rapid extraction and depletion of minerals (or other nonrenewable raw materials) that will occur under the present custom of unrestricted mining, use, and throw away. Some key metals will be 'mined out' before the year 2000 if this pattern persists. B. Depletion time can be extended by partial recycle and less wasteful use. C. Efficient recycle, combined with stringent conservation and substitutions (using more abundant alternative material wherever possible), can extend mineral depletion curves indefinitely. (Diagram adapted from Cloud, 1969.)

*Extract from Odum, E.P.: *Fundamentals of Ecology*, 3rd edition. Published by W.B. Saunders Company, Philadelphia, 1971.

resources) to anywhere near the level of the United States, severe shortages would develop tomorrow. In developed countries the per capita demands for a moderately scarce metal such as copper is projected to triple by the year 2000! Aluminium is cited by Cloud as an example of the general situation because it is not, relatively speaking a scarce metal. Prior to 1945, the United States produced most of the ore (bauxite) that it used, but by 1960 this country was importing three times as much ore as it was mining on its own lands. It is obvious that we can no longer afford 'one-way' aluminium beer and soft-drink cans (or similar 'uncycled' uses); we have to substitute or recycle, or both. The industrialized countries, in general, are no longer self-sufficient in either minerals or fossil fuels; they depend more and more on exploitation of these natural resources from the undeveloped part of the globe, where, of course, the supply is finite, cannot be increased, and, in fact, will decrease as these countries start to use their own wealth.

The other concept introduced by Cloud is the graphic model of *depletion curves*, as shown in Figure 1. With the present procedure of 'mine, use, and throw away', a huge boom and bust is projected, as shown in curve A. The time scale is uncertain because of lack of data, but the 'bust' could begin within this century since certain key metals such as zinc, tin, lead (needed for that electric car!), copper, and other metals could be mined out in 20 years insofar as the readily exploitable reserves are concerned. Likewise, fuels such as uranium-235 and natural gas could also be gone by then. If a program of mineral conservation involving restrictions, substitutions (using less scarce minerals wherever possible), and partial recycle were to be started now, the depletion curve could be flattened as shown in curve B. Efficient recycling combined with stringent conservation and a reduction in per capita use ('power down' by developed countries) could prolong depletion for a long time, as shown in curve C. It should be noted that even with perfect recycle depletion would still occur. Thus, if we were able to recycle 60 million tons of iron each year, about half a million tons of new iron would be needed to replace the inevitable loss from friction, rust, and so on.

Inventories and projections regarding the mineral fuels are firm and generally agreed upon. . . . Pollution rather than supply will be the limiting factor for industrial energy. As already indicated, natural gas and uranium will be soon gone, but oil and coal will last longer. In the meantime 'breeder' reactors and possibly the development of fusion atomic energy should fill the energy gap. Thus, for the time being at least, biotic and mineral resources are more critical than energy; it can be hoped that these limits will actually prevent man from attempting to increase energy use to the point of literally burning up the world.

Recommended reading is the slim but potent volume entitled *Resources and Man*, published in 1969, as the summary report of a committee of the National Academy of Science (Preston Cloud, Chairman). The report urges caution in relying on the optimistic projections of some technologists in regard to: (1) the sea as an unlimited supply depot, and (2) the extraction of metals from lean ores with the use of vast quantities of cheap atomic energy. As we have already noted most mineral wealth (and exploitable food) is located near the shore and does not by any stretch of the imagination provide anything more than a supplement to the continental supplies. The use of vast amounts of atomic energy to extract low grade ores would convert the world into one vast strip mine and create hazardous and expensive waste disposal and pollution problems, which we hope will never have to be evaluated! The National Academy report ends with 26 recommendations that all boil down to a twofold theme: *both human population control and better resource management that includes recycle are needed — NOW.*

Question

Why do we need to have both human population control and better resource management?

4.4.5. Population explosion*

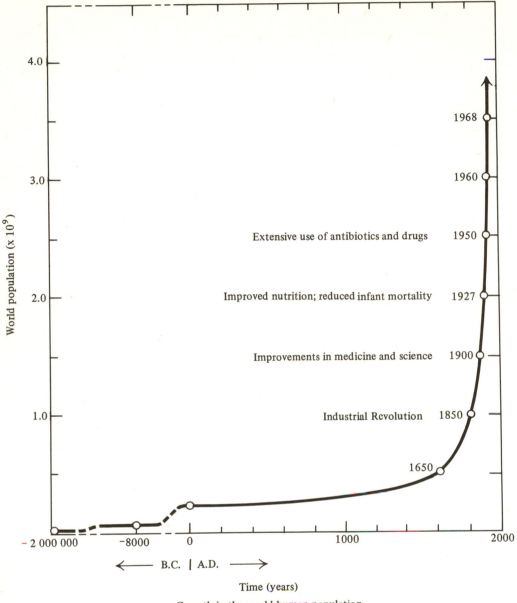

Growth in the world human population

In spite of a great deal of publicity on the subject, most of the earth's people are not aware of its significance and urgency. Even in the presumably well-informed United States, a Gallup poll taken in the mid-1960's indicated that two-fifths of Americans did not view uncontrolled population increase as a major world problem.

Most animals have built-in regulatory mechanisms to limit population size. These by no means would be acceptable to humans: release, when there is crowding, of noxious substances that keep down mating frequency; hormonal inhibition of reproduction induced by the social stresses of crowding; spread of genes causing death or sterility; and so forth. Although a human female normally can produce no more than 10 or 12 offspring, which seem remark-

*Extract from *Heredity, Evolution, and Society*, I. Michael Lerner. W.H. Freeman & Co. Copyright © 1968.

ably few when we learn that the tapeworm can lay 120 000 fertile eggs a day, man is increasing his numbers at a rate that is probably greater than that of any other organism. Diseases and famine, pestilence and war, infanticide and human sacrifice, unpleasant as it may be to admit it, have regulated population growth hitherto. They no longer do so.

Years ago	Cultural stage	Area populated	Assumed density per square kilometre	Total population (millions)
1 000 000	Old Paleolithic		0.00425	0.125
300 000	Middle Paleolithic		0.012	1
25 000	New Paleolithic		0.04	3.34
10 000	Mesolithic		0.04	5.32
6000	Village farming and early urban		1.0 / 0.04	86.5
2000	Village farming and urban		1.0	133
310	Farming and industrial		3.7	545
210	Farming and industrial		4.9	728
160	Farming and industrial		6.2	906
60	Farming and industrial		11.0	1610
10	Farming and industrial		16.4	2400
A.D. 2000	Farming and industrial		46.0	6270

Fig. 1. Estimated population size and density from the Paleolithic Age to the end of this century. (From Edward S. Deevey, Jr., *The Human Population.* Copyright © 1960 by Scientific American, Inc. All rights reserved.)

A look at some statistics may help us see the problem, but a word of caution is necessary. All the demographic data to be cited are only coarse approximations. Census figures are notoriously unreliable, and estimates of population size and growth are subject to gross errors. For instance, figures given for the population of Communist China vary at their extremes by 100 per cent. The population of Mauretania in Africa is estimated by its government as being threefold the figure given by the United Nations. During the partition of Pakistan and India, the number of migrants from one country to another was computed by dividing the amount of salt issued to them by the average salt requirement. The 1960 census of the United States in which the data were analyzed by the latest computerized techniques, showed that there were in the United States, 1670 fourteen-year-old widowers, a figure of doubtful validity.

But no matter how much we allow for such inaccuracies, the picture is clear. The history of population growth in this world is shown in Figure 1. Put in another way, the tremendous acceleration in the increase of the human population is illustrated by the following figures: it took 4.5 billion years for the population to reach one billion* people (1830); it took 100 years for it to reach two billion (1930); it took 30 years to reach three billion (1960); at the present rate of increase it will take 15 years to reach four billion (1975). About 5 per cent of all *Homo sapiens* that ever lived roam the earth today.

We can calculate, as an average, that over the whole period of human history, the population doubled every 70 000 years. As recently as 1650 the rate of population growth required 200 years to increase the number of people on earth twofold. Today, the doubling time is about 35 years, varying from 23 years in Brazil to 44 in the US and 76 in Japan. According to these statistics, the 15 million people killed in the battles of World War II were replaced in something like three and a half months. The earth's population today stands at over 3.2 billion; the annual rate of increase is 2.5 per cent or 80 million people. Every day the population rises by 220 thousand; every hour by more than 9,000.

How are these people to be fed? Today there are 1.2 acres of cultivated land per person, and yet 60 per cent of the people have less than the average requirement of 2200 calories a day. By the year 2000, through alienation of land and population growth, six billion people will have less than half of this amount per person available to them. In the next third of a century, we shall have to triple our food output, increase our production of lumber threefold and our output of energy, iron ore, and aluminium fivefold, just to permit the growing number of human beings to reach a decent living standard.

Technological advances in food production might alleviate the problem of feeding the growing world population. Some authorities think that the earth is potentially capable of supporting 50–100 billion people. But at present our food production is growing too slowly. Just to keep pace with the present population growth, without attempting to relieve widespread under-nourishment, an increase of 2.5 per cent a year in food production is needed. Our recent increases have been on the order of one per cent a year. Thus, while North Americans produce more calories than they need, the average food supply available in India and elsewhere is lower than it was before World War II.

Many ways of resolving the situation have been suggested. The lack of fresh water for irrigation is one of the limitations on food production. Increased use of rational and intensive methods of farming, making fresh water from the sea, heating lakes to encourage cloud formation, and fish and

*1 billion = 10^9

game farming are all possible techniques for augmenting food sources now available. New forms of food may be produced. Man is a heterotroph, relying on the photosynthetic activity of plants and protists for energy-binding. This creates a rather inefficient food chain, because much energy is lost in the process of conversion. Although selective breeding of plants and animals has reduced the amount of raw material needed to manufacture foodstuffs, 80—90 per cent of the available energy is still dissipated in every link of the chain. Perhaps the longest food chain of all appears in the diet of Eskimos. An Eskimo must eat five pounds of seal to gain a pound of weight, each pound of seal being derived from five pounds of fish, a pound of which is manufactured from five pounds of shrimp or other invertebrates, each pound of which takes five pounds of algae to produce. In sum, it takes 625 pounds of algae to make one pound of Eskimo, with at least 99.84 per cent of the original energy bound by the algae being lost in the process.

One method of shortening the food chain is to grow for direct human consumption plants that would satisfy nutritional requirements presently met by eating animals. For instance, it is possible to manufacture edible protein from otherwise inedible leaves. Similarly, the use of marine protists or yeasts for food is a possibility but the deeply ingrained preferences people hold for certain food makes utilization of such possibilities difficult.

Another complication is the fact that most plant proteins are deficient in certain amino acids essential for the synthesis of proteins in human cells, which lack the appropriate enzymes to manufacture them. Hence, supplementation of a plant diet with synthetically produced amino acids or development of strains of plant food with higher levels of these amino acids may be needed. At least two mutations in corn that increase the amounts of the amino acids lysine, tryptophan, and methionine have already been found. These three substances are among those that man needs and obtains, at present, mostly from animal proteins. Methods using microorganisms to produce high-grade protein from such nonliving materials as petroleum are also being developed. The ultimate technology of food production may depend entirely on synthetic processes, although human dietary habits and palatability standards may have to be modified. Perhaps the need for these adjustments will be circumvented, however, by the manufacture of synthetic material indistinguishable from familiar foodstuffs.

The costs, of course, will be enormous. It is estimated that to feed the population of Southeast Asia alone in 1980 more than 22 billion pounds of bulk protein would have to be produced. At present prices this would cost about 55 billion dollars a year.

In any case, the present doubling rate of the population is such that even if the nutritional deficit could be erased completely, other problems raised by overpopulations would not be solved. Various kinds of limits to the number of people on earth have been imagined, including, for instance, eventual scarcity of nitrogen, which is necessary to make living beings, although this is unlikely. Perhaps, the grimmest picture, relieved by a light touch, has been painted by the British physicist J.H. Fremlin.

He has suggested that the absolute limit to the human population on this planet will be determined by terrestrial overheating. People and their activities generate heat that must be dissipated. If the population increased to one quadrillion people, temperatures throughout the world would reach those currently known in equatorial areas. Adequate cooling devices would make it possible to have 60 quadrillion people on earth. The population density would then be 120 million per square kilometer; today it is 18. When the population rose to 10 quintillion, overheating would literally cook people. At the present rate of population growth, this would happen in about 900 years.

Fremlin imagines many difficulties even if the population stabilized at 60 quadrillion, for example, the housing problem. Perhaps technological advances would permit us to build 2000-story buildings, covering both land and sea, with 1000 stories of each housing food-production and refrigeration machinery. Of the rest of the space half would be occupied by wiring, piping, ducting, and elevators, leaving 3¾ square meters of living space per person. Food would have to be liquid, clothes would not be worn, cadavers would be immediately processed into food, and each area of a few square kilometers, containing several billion people, would have to be nearly self-sufficient.

Very little movement of people would be tolerated, but still 'each individual could choose his friends out of some ten million people, giving adequate social variety', and world-wide television of unexcelled quality would be available. With the present size of the human population, the birth of a Shakespeare is an exceedingly rare event. In the world of 60 quadrillion some ten million Shakespeares ('and rather more Beatles') might be expected to be alive at any given time.

This apocalyptic vision is not likely to become reality. Our problem is how to put a brake on the increase in the number of people on earth before it does. The main causes of the population explosion are considered in Box A. In brief, they lie in the exponential (Malthusian) nature of population growth and in the increase, owing to advances in sanitation and medicine, in the percentage of human beings who survive to an age at which they can produce offspring.

Many suggestions have been made as to how population increases can be controlled in addition to the very obvious method of contraception. Ireland's population growth has been limited by a traditional prolongation of the period of celibacy, a method that seems increasingly unlikely to catch on elsewhere in the world. Japan has slowed down its population explosion by encouraging abortion. Legalizing abortion, although it is meeting some religious opposition in the United States, seems a sensible move, but the medical and psychological effects of requiring repeated and numerous abortions for the same woman may rule out its value as a method of population control. Sterilization has been suggested as still another means. Yet it has been computed that the scale on which it would have to be undertaken is not a practical one. For instance, in India, the birth rate (42 per thousand) and the death rate (19 per thousand) might be brought into equilibrium at 16 per thousand by 1991, with the population stabilized at 657 million (it was 438 million in 1961). But this would have required an increasing rate of sterilization, with everybody above the age of 22 having been sterilized by the time the equilibrium were reached. And if the population were to be reduced, the extent of sterilization would have to be even greater. Reduction of population size by any humane means is in general a very slow process. For example, to reduce the present population of Britain to that of 1910 by any reasonable methods might take 200 years.

Even the suggestion that encouragement of homosexuality could keep down population size has been made. Now, so far, no clear-cut genetic basis for homosexuality, or other types of sexual gratification that do not result in conception, has been demonstrated. Should there be one, obviously in the long run this method of population control would be self-defeating: natural

127

Box A. Some causes of the population explosion

A few data gathered from various sources will illustrate some of the reasons for the current population explosion. To start with, if the birth rate is greater than the death rate, population size will increase exponentially, so long as the difference is greater than zero. Other factors also increase the rate of population growth. A minor one is the current tendency to marry younger. In Europe, the average age at which persons enter into marriage contracts is 0.5—2.8 years younger than it was two or three decades ago. The effects of this change are shown in the following table, which gives the average number of children in a sample of families.

Socioeconomic status	Age at first marriage	
	20-24	30-34
Professional	1.8	1.1
Skilled labor	2.3	1.1
Unskilled labor	3.1	1.3

Of more significant effect are the great increases in life-span in this century, largely owing to the rapid progress in prophylactic and therapeutic methods of controlling infectious disease.

Life expectancies in different periods of history have been estimated as follows:

Period	Life-span in years
8000—3000 B.C.	18.0
(Stone and Bronze ages)	
A.D. 800—1200	31.0
1600—1700	33.5
1800—1900	37.0
early 1900's	57.4
mid-20th century	66.5

Thus, only 11.8 per cent of dated Paleolithic skeletons represent human beings who lived more than 41 years; today in the United States, 95 per cent of the people survive past 40. Some examples of recent changes in life expectancy may be instructive:

Country	Period	
England—Wales	1910—12	1958
	53 years	71 years
USSR	1896—97	1955—56
	32 years	66 years
Jamaica	1919—21	1950—52
	38 years	57 years

Especially dramatic changes have been observed in the underdeveloped countries because of the introduction of sulfa-drugs, antibiotics, DDT, and other means of controlling the spread of infections. At the present time, 60 per cent of the world's adults — but 80 per cent of the world's children — live in underdeveloped areas. Examples of lowered infant mortality are given by the table below, which shows the percentage of infants surviving a year after birth:

Place	Period	
London, England	1860	1960
	70.0	98.0
India	1901	1962
	77.0	90.5

One evolutionary effect of the increased probability of surviving until the reproductive age is that, whereas survival *per se* was an important concomitant of Darwinian fitness in the past, today, with the high level of survival, most of the selection in human population is based on differential fertility. As voluntary contraception becomes more widespread and standards of living improve, philoprogenitiveness may become the main selective agent. In other words, natural selection will lead to an increase in the proportion of the more prolific parents, a fact that may eventually negate the population-regulating efficacy of birth control.

selection will tend to weed out genotypes for preferences for sexual behavior not leading to production of offspring (see the remarks on philoprogenitiveness in Box A).

Visionaries who suggest that the problem of overpopulation can be solved by exporting mankind to other planets can hardly take comfort in the fact that it would take only 50 years at the present rate of population growth to bring Venus, Mercury, Mars, the moon, and the satellites of Jupiter and Saturn to the same population density as the earth. Colonization of Saturn and Uranus would give only another 200 years' breathing space. Besides, going outside of our solar system presents what now seem to be insurmountable problems of time and cost.

So contraception seems to be the logical method to use. Cheap and efficient techniques are already available, and only ignorance, religious dogma, and socio-political considerations seem to stand in the way of universal acceptance of family planning. Opposition from organized religious groups appears to be gradually crumbling, and ignorance is being dispelled by education, but other problems still exist. For example, advocacy of limiting the number of children per family in minority groups sounds hollow indeed, when wealthy men raise huge families.

Among those scholars who attempt to foresee the future, there are some who believe that man faces an inescapable dilemma because of the rivalry among nations. If, on the one hand, a given country pursues a vigorous policy of limiting the growth of its population, it will (other things being equal) incur a political disadvantage vis-a-vis a country that does not: on the other, lack of restraint will increase the intensity of the struggle among nations for resources. Others, less bleak in their forecasts, hope for an eventual world-wide society — whether a totalitarian state or a free association is envisaged depends on the optimism of the forecaster — without national boundaries and with demographic policies that may make this world a good place to live in. It is possible that in such a society sterilizing chemicals will be added to drinking water, and conceptions will be the result of a premeditated act,* rather than occurring by default, as most do now. This could imply the licensing of reproduction, an uninviting prospect, but, in the view of some demographers, an unavoidable one.

---

*As a consequence, one of the best known nursery rhymes may have to be revised:
Mother, may I conceive a child?
Yes, my darling daughter.
Hang your clothes on a hickory limb,
But don't go near the water.

129

4.4.6. Stabilization of toxic mine wastes by the use of tolerant plant populations*

R.A.H. Smith
Formerly Research Assistant, Department of Botany, University of Liverpool (now The Nature Conservancy, Merlewood Research Station, Grange-over-Sands, Lancashire)
A.D. Bradshaw
Professor of Botany, University of Liverpool

Synopsis
There is a need to find an economic and permanent method of stabilizing mine wastes containing toxic levels of metals — both to prevent pollution and to improve the visual appearance of mine workings. It should not involve the use of topsoil, since this is expensive and usually not available. Naturally occurring metal-tolerant populations of wild plant species offer a possible technique, since their metal tolerance should overcome the metal toxicity which is the main factor causing mine wastes to be devoid of vegetation. The other factor, lack of plant nutrients, can be overcome by the use of fertilizer.

Investigations on mine wastes in various parts of England and Wales, of a large number of such populations, show that, provided that fertilizer is given, they will grow very successfully. Ordinary commercial material of the same species may grow for a short time, but it dies within a year. The metal-tolerant material roots very well and is, therefore, much less susceptible to the drought which can occur in coarser grades of waste. Metal tolerance in populations is specific to individual metals, and different species themselves have preference for particular calcium levels and acidity.

As an outcome, three different types of material of the grass species *Agrostis* and *Festuca* are being multiplied to be available for the stabilization of different types of waste in temperate climates. Since metal tolerance can also be found in species in tropical vegetation, material could readily be developed to cope with mine wastes in these areas also.

Mining for metals was widespread in Britain from the sixteenth century until the end of the last century; this industry has left a legacy of disused and unsightly waste heaps. The smelting industry has also left similar tips. Now at the present day mining and extraction may commence again. Elsewhere in the world mining has gone on continuously, leaving piles of waste material.

Table 1. *Analyses of waste dump materials (total p.p.m.)*

Site	County	Ca	K	P	N	Cu	Pb	Zn
Trelogan	Flintshire	74 560	503	401	200	83	10 510	35 940
Halkyn	Flintshire	310 500	908	290	100	59	7 400	11 680
Parc	Caernarvonshire	35 700	783	137	200	43	1 750	9 450
Darley	Derbyshire	226 250	1190	105	200	131	11 100	34 750
Snailbeach	Shropshire	236 880	211	110	100	25	20 900	20 460
Ecton	Staffordshire	137 120	825	116	1100	15 350	29 850	20 150
Nenthead	Cumberland	28 750	3160	74	500	65	2 020	10 540
Goginan	Cardiganshire	7	320	203	200	200	15 400	1 370
Van	Montgomeryshire	5	457	125	900	93	40 500	1 200
Glenridding	Westmorland	3 830	862	201	200	40	5 650	2 290

These waste heaps are usually more or less bare of vegetation cover. This is due to several factors, the most important of which is probably the residual high level of heavy metals such as lead, zinc and copper (see Table 1), which are extremely toxic to plant growth. Other factors are macronutrient deficiency, unfavourable physical soil structure, surface movement and dryness. The heavy metal toxicity has a very specific effect, causing more or less complete inhibition of root growth. This increases the effect of dry conditions, lack of roots making the plants much more susceptible to drought.

In Britain these heaps are often situated in areas of great natural beauty, such as the Lake District, the Peak District, Cornwall and Wales. Rehabilitation of these sites is therefore particularly important for purely aesthetic

*Reprinted from *Trans. Institute of Mining and Metallurgy*, 81 (1972), by permission.

Fig. 1. View of lower part of Trelogan mine showing spread into adjacent farmland due to movements by wind and water: original experimental material within enclosure; further plantings for seed production in distance.

reasons, but it is also important in order to prevent heavy metal pollution, which may have serious effects on crops and public health. Future mining may not be permitted if proper methods for dealing with tailings and other waste material are not found. Pollution can result from dust blow, washout into rivers of mobile heavy metal ions, or large-scale erosion of material. Dust blow on a large scale occurs, for instance, from Frongoch lead–zinc mine in mid-Wales and from Trelogan lead–zinc mine in Flintshire (Fig. 1).

In the Conway Valley in North Wales a washout from Parc lead–zinc mine tailings dump in 1964 caused the ruination of a large area of fertile lowland pasture, which was covered with up to 0.5 m of spoil. When it is considered that this sort of material may contain up to 80 000 p.p.m. lead and 60 000 p.p.m. zinc, and that in an average soil, levels in excess of 500 p.p.m. are very toxic to vegetation and stock, the seriousness of the pollution problem will be appreciated.

There would seem to be two alternative methods of dealing with this toxic waste: removal or reclamation.

Removal for use as grit for roads or as a base for construction is a possibility in Britain, where the amount of material is not great relative to the density of the population. But the former use, in particular, which is occurring, for instance, at Trelogan lead–zinc mine, Flintshire, poses the danger of spreading the pollution farther afield. It also seems unlikely that these rather limited uses will make much impression on the vast volume of material left from the industry, particularly in distant parts of the British countryside, and in new large-scale operations elsewhere in the world.

Reclamation by covering with vegetation thus seems the only possible alternative, although stabilization by resin compounds and other physical means may be a temporary solution. We have been financed by the Natural Environment Research Council to investigate the possibilities of developing a cheap and practicable reclamation scheme for these sites.

The dumps vary considerably in several important ways. First, the metal content can vary widely from one part of a heap to another. The metals present can also vary: lead and zinc are usually found together and copper separately. But all three can be associated in different ways; thus, at Ecton, Staffordshire, all three occur together, and in one part of Parys Mountain, Anglesey, copper occurs by itself, and in another part together with lead.

The associated minerals vary enormously. The most important variation is the presence or absence of calcium, nearly always as calcium carbonate. This radically alters pH and, hence, the environment of the waste as a medium for plant growth.

Finally, because of different extractive systems, the material may be left in coarse lumps or as very fine tailings. The coarse material is not at all retentive of water, although fine tailings may have excellent water-retaining ability. The topography of the dumps is also very variable.

There have been several attempts in the past to reclaim these toxic waste tips. These have frequently involved partial turfing, in the hope that the vegetation from the turves would colonize the intervening bare areas. At Nenthead lead–zinc mine, Cumberland, this technique has been tried with limited success. At Nant, Caernarvonshire, on the tailings from Parc lead–zinc mine, instead of colonization between the turves there has been erosion of the intervening material to a depth of 20 cm.

An alternative technique is to blanket the waste with a layer of non-toxic material, such as soil, sewage sludge or domestic refuse, and sow seed on this. This has been extensively explored at Swansea for smelter waste with considerable success.[6] This method has also been used with reasonable success at Glenridding lead–zinc mine in the Lake District, where, in 1953, seed was sown on a 10 cm layer of sewage sludge. But although the scheme was quite successful in the beginning, erosion has now been initiated because of overgrazing by sheep and the dieback of vegetation caused by the spread of toxicity to the surface. Once started, this erosion continues since the plants tend only to root in the sewage, which forms a skin which is more or less unattached to the underlying spoil.

In any event, it must be remembered that, because of high transport charges, even over short distances, methods involving covering the spoil with a non-toxic layer are extremely expensive, usually costing more than £300 per acre. Topsoil, etc., is often not available near to these sites, and in these cases the outlay per acre can be more than £1000.

If we are to find a cheaper method, we should perhaps ask what precisely is achieved by a blanketing layer and whether there are specific alternatives. A blanketing layer has three main properties. First, it provides a water-retaining covering: but it is questionable whether in humid climates, such as that in Britain, water lack is a limiting factor in mine wastes except in very extreme situations. Secondly, it provides an adequate supply of nutrients: but it could be cheaper and more efficient to supply nutrients by direct fertilizer application. Thirdly, it overcomes toxicity: but evidence from Glenridding and the Lower Swansea Valley indicates that this may not be achieved on a long-term basis because the toxic metals move up from below.

What then is an alternative?

It has been noticed for a long time that most of these tips are naturally sparsely colonized by a very characteristic and limited group of species, mainly grasses. On acid waste in Britain *Agrostis tenuis* and *Festuca ovina* predominate. On calcareous material the dominants are *Agrostis stolonifera*, *Festuca rubra* and *Deschampsia cespitosa*. In recent years many of these species have been shown to possess populations which are tolerant of the high heavy metal contents and low macronutrient levels present in the waste.[2, 8, 10] The heavy metal tolerance is the result of the metal ions being rendered immobile and innocuous by complexing in the root cell walls.[9] The tolerance is specific to individual metals.[5]

The unique ability of these few species to colonize such toxic sites is believed to be a result of their unusual capacity to evolve populations possessing tolerance to heavy metals. It would thus seem obvious that the most satisfactory long-term solution to this problem would be to use these tolerant populations, which are already adapted to growing on these difficult habitats. Experiments at Swansea on toxic smelter waste with vegetative plant material gave the clear suggestion that the technique had considerable potential.[4] It was on this assumption that our reclamation scheme was formulated, the aim being to use only (1) seed of tolerant populations and (2) fertilizer amelioration. When used on a large scale, seed would be more practicable than vegetative material. The fertilizer would be necessary to overcome the major plant nutrient deficiencies found in mine wastes.

Accordingly, a series of experimental trials was begun, initially under greenhouse conditions and then in the more exacting conditions of the field, all the species mentioned above being used.

Experiments
The preliminary greenhouse trials were carried out over the winter of 1968 on a range of acid and calcareous lead–zinc mine wastes from Wales with various races of mine and normal populations of a number of species with different fertilizer ameliorations. The choice of which species to use throughout these trials was determined by what the natural species were on the site in question. The outcome of these trials is exemplified in Fig. 2 [overpage], which shows the rooting of the different populations in waste from Trelogan lead–zinc mine, Flintshire. The tolerant populations were superior throughout the experiment. Even over the short time the experiments ran there was a clear improvement in total amount of growth due to fertilizer, especially nitrogen and phosphorus in combination, although organic manure was even more effective, as could be expected.

These preliminary experiments gave some indication of the value of the use of tolerant seed for reclamation of these sites, and also some idea of the fertilizer levels to use. The real test, however, was to apply these to field conditions and see if the successes were repeated.

Trials were sown in the spring of 1969 on four mine sites in Wales — three calcareous and one acid. Growth of these plots was highly satisfactory. The superiority of the tolerant populations was more marked in field conditions than in the greenhouse: this is probably due to the effect of toxicity on root growth, causing the drought conditions which followed sowing to differentiate very markedly between tolerant and non-tolerant populations. The important effect of fertilizer treatment was apparent on all plots, slow-release fertilizer (John Innes base) being usually a little better than an ordinary mineral mixture of NPK. This is clear from the dry weight data for the plots at Trelogan after seven months' growth (Fig. 3, [p. 135]). Other

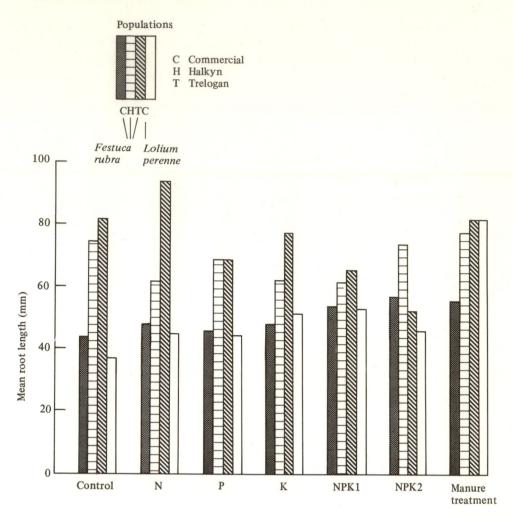

Fig. 2. Root growth of different populations on waste from Trelogan mine in preliminary greenhouse experiment with different fertilizer treatments.

plots behaved similarly, although on the acid site it was the tolerant populations of species adapted to acid conditions which did best.

The history of the experiment illustrated by the Trelogan plot is interesting (Fig. 4). To begin with, everything grew well, particularly in the fertilized plots, owing to the complexing of the toxic metals by the fertilizer (phosphate, for instance, renders heavy metals very insoluble). After six months, however, the superiority of the tolerant material was becoming clear, although the rye grass *Lolium perenne*, a normal agricultural grass chosen to be a general control, was apparently doing well and the non-tolerant populations of the other species were still surviving. In a practical situation it could have been presumed that rye grass was a real alternative to tolerant material. The next twelve months, however, showed how erroneous this conclusion would be. For rye grass was totally dead by the end of this time and the non-tolerant populations of the other species almost so. In one way or another this was repeated in the other sites. The difference in the behaviour of tolerant and non-tolerant *Festuca rubra* at Trelogan after three years is shown in Fig. 5 [p. 137].

This technique was thus demonstrated to be successful in the long term under field conditions, but only on a limited number of sites, which, although they had been initially selected for their observed intractability, were not necessarily those with the highest metal contents and potential toxicity.

134

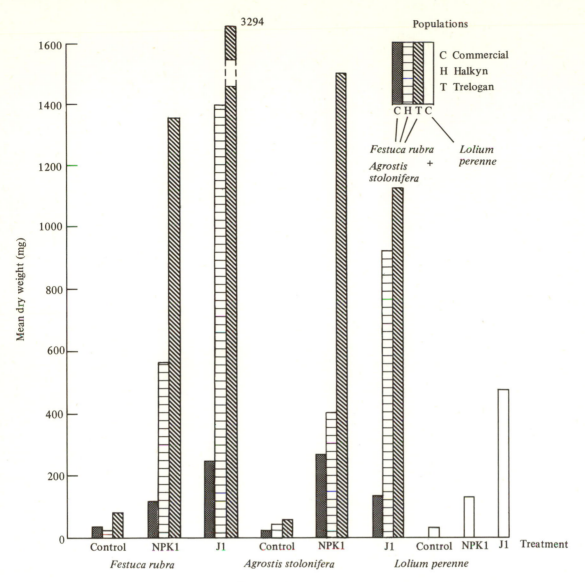

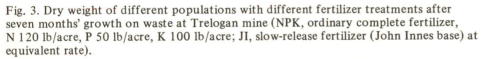

Fig. 3. Dry weight of different populations with different fertilizer treatments after seven months' growth on waste at Trelogan mine (NPK, ordinary complete fertilizer, N 120 lb/acre, P 50 lb/acre, K 100 lb/acre; JI, slow-release fertilizer (John Innes base) at equivalent rate).

So trials were set up on the original sites and on a further six sites, with the aim of testing the applicability of the one scheme on a number of diverse areas. These trials were sown in the autumn of 1969. Analyses of the material from all these sites are shown in Table 1 [p. 130]. Obviously there is an enormous range of contents of metals and nutrients, particularly calcium.

On each site was sown a commercial race, the tolerant race from the particular site, a tolerant race from a standard site, and also commercial *Lolium perenne*. The species sown depended on what was naturally growing on the site.

As an example of the sort of thing that happened, dry weight yields for different populations of *Festuca rubra* and for *Lolium perenne* after a year's growth on the calcareous plots with a slow-release complete fertilizer (John Innes base) treatment are given in Fig. 6 [p. 138]. On the majority of sites both tolerant races were considerably superior to the non-tolerant. On sites such as Nenthead and Halkyn sheep grazing occurred, which more or less eliminated this differentiation since, in addition to reducing the overall

135

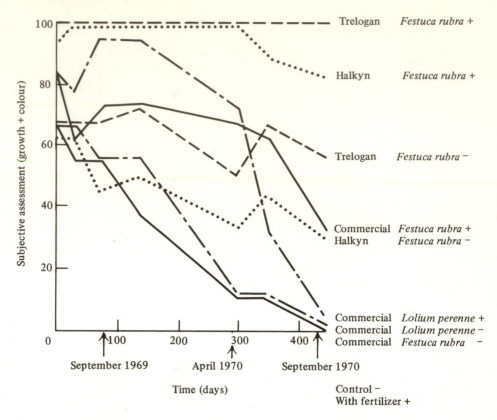

Fig. 4. Visual assessment of different populations, with and without fertilizer, over a period of two years on waste at Trelogan mine.

growth and, therefore, differences between the races, the grazing was selective for the greener more conspicuous tolerant material. The Trelogan race was superior to the native race on most sites; this is perhaps because the metal contents at Trelogan are very high for both lead and zinc and, therefore, this population will be highly adapted to cope with the levels of these metals present in the other sites. This was true for both *Festuca rubra* and for *Agrostis stolonifera*.

It thus seems that it would be possible to use the same seed population to reclaim a range of sites, provided that these were of similar acidity. This finding is very important since it means that only a limited number of seed races need be developed. On calcareous sites the species with successful populations were *Festuca rubra* and *Agrostis stolonifera*. On acidic sites the species were *Agrostis tenuis* and *Festuca ovina*.

Despite a dry spell immediately following sowing, establishment of the initial, 1968, spring-sown plots was very satisfactory — indicating that drought was not a limiting factor. Plots sown the following spring, an exceptionally dry one, still showed reasonable establishment, although some germination was delayed until after the drought. So, except in freak conditions, germination and establishment should be satisfactory in spring, and this appears to be the best time for sowing. Autumn-sown plots, although germinating well, make very slow growth during the winter; by spring, leaching of the fertilizer causes their response to the favourable conditions to be inferior to that of the spring-sown trials. Plots approaching a commercially practical size sown in the spring of 1971 and subject to severe drought have proved equally successful (Fig. 7, [p. 140]).

A series of small plots was also sown on a variety of steeply sloping sites

Fig. 5. Close-up of difference in performance of Trelogan non-tolerant (*left*) and Trelogan tolerant (*right*) populations of *Festuca rubra* on waste at Trelogan mine after three years: fertilizer was added at planting only.

on the sides of lead—zinc waste dumps. In these the superiority of the Trelogan *Festuca rubra* population was complete: in many sites it was the only population to establish. Having established, it then appeared to grow as well as if it were on a flat site. Under these conditions *Agrostis stolonifera* did not do well: it has very small seeds and, hence, small and vulnerable seedlings.

There are certainly no signs of drought problems anywhere once the seedlings are established. This is almost certainly due to the fact that mine tailings are always finely ground and, therefore, have adequate moisture retention in the climate of the British Isles. But it is also due to the fact that tolerant material with adequate fertilizer produces a large deeply penetrating root system, which can tap water in the deeper parts of the dumps. It is noticeable that actual cause of death in non-tolerant material is usually drought, because of its shallow, inadequate, root system.

The technique is therefore effective in a very wide variety of conditions. The only drawback is that there does not seem to be any possibility of a direct economic return. Use of the vegetation for grazing is not feasible since, although much reduced in the tolerant races, appreciable amounts of toxic metals enter the plant (Table 2 [p. 139]). In long-established swards

137

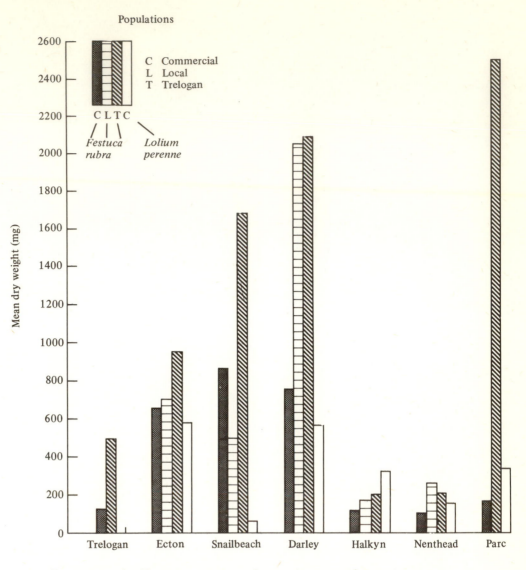

Fig. 6. Dry weight of different populations after twelve months' growth on seven different calcareous lead—zinc mines with fertilizer (local population is derived from the site where trial was carried out).

the metal contents may rise even higher — to about 0.4 per cent. It seems likely that reclamation will be for the purposes of amenity improvement and reduction of pollution, which are, of course, crucial, rather than for agricultural purposes.

It would be very valuable if similar tolerance could be found in tree species, but populations of trees growing on toxic mine waste are extremely uncommon and do not give convincing evidence of tolerance. It was hoped, however, that, since even in commercial grass races there is the occurrence of occasional tolerant individuals,[2] the same might be true for trees. Accordingly, screening of four species of conifer is being carried out by growing them, from seed, on mine soil. After eighteen months quite a number of individuals are still surviving, particularly of Scots Pine, but growth is so much reduced over that in control soil that it is doubted if any individuals exist which are truly tolerant.

Conclusion
The use of tolerant plant populations and fertilizer would seem to be the

138

only reclamation technique, not involving topsoil or equivalent material, to give reasonable prospect of long-term success on these very intractable sites. At present the ideal fertilizer treatment would be use of John Innes base slow-release fertilizer at a rate of 20 cwt/acre, but we have found that a less expensive and almost as efficient alternative is a standard 15 : 15 : 15 NPK fertilizer at 7 cwt/acre. Cost—benefit trials are continuing, however, with a range of products, including slow-acting materials, and it may be that a better alternative will be found. Expensive repetition of fertilizer applications will probably be unnecessary since these metal-tolerant populations are also adapted to low nutrient levels.[7]

Table 2. *Analyses of metal contents of different populations growing on waste at Trelogan (total p.p.m. in shoots)*

Species	Population	Lead		Zinc	
		Without fertilizer	With fertilizer	Without fertilizer	With fertilizer
Agrostis stolonifera	Commercial*	1140	810	2965	1930
	Halkyn	700	454	1783	1310
	Trelogan	462	623	1446	1528
Festuca rubra	Commercial*	2124	495	3854	1915
	Halkyn	102	979	2483	2800
	Trelogan	397	808	1545	1975
Lolium perenne	Commercial†	2293	850	4504	2500

*Plants almost dead
†Plants died later

Throughout most of the trials seed was sown at a rate of 200 lb/acre but this is very high from an agricultural point of view. It seems likely from further trials that this can be reduced to 10 lb/acre or less of tolerant material if a nurse grass is added, without reducing ultimate vegetation cover appreciably.

This reclamation technique will be cheaper than conventional methods with soil. Inorganic fertilizer and cultivation costs, excluding regrading, should not exceed £40 per acre, and the seed, if it were developed commercially for about £2/lb, would not inflate the total cost to more than £80—£100 per acre — a small fraction of the outlay with conventional methods. If the topography of a site demands the use of hydraulic seeding methods, tolerant material will be absolutely invaluable since this method cannot possibly remove metal toxicity either temporarily or permanently. A trial recently carried out at Tynagh mine, Eire, on lead—zinc waste with commercial and Trelogan *Festuca rubra* with hydraulic seeding is very promising.

Since tolerance is specific to individual metals, it is necessary to have available populations with appropriate tolerances. At the same time species must be chosen which are adapted to the acidity or alkalinity of the site. Thus, for British and other equivalent conditions the following grass material is being sent for seed multiplication to be available on a commercial scale:

Festuca rubra   Red fescue, for calcareous lead—zinc contaminated wastes

Agrostis tenuis   Bent grass, for acidic lead—zinc contaminated wastes

Agrostis tenuis   Bent grass, for copper-contaminated wastes

Fig. 7. View of large-scale experiment at Trelogan with tolerant *Festuca rubra* and *Agrostis stolonifera* (note almost complete absence of any other plant growth).

Although *Agrostis stolonifera* has excellent populations for calcareous lead–zinc contaminated wastes, its seed production is poor and its seedlings rather vulnerable to drought. It would, however, be very suitable for planting by hand in damp situations, since it spreads extremely readily by stolons.

In other areas of the world it is known that natural metal-tolerant populations of other species exist, and tolerance to such metals as chromium and nickel can be found.[1] Some of these could be developed very easily to be used in the climatic regions to which they are adapted. The authors would be very pleased to cooperate with anyone interested in such developments.

If legislation is introduced to reduce the environmental impact of metal-liferous mining, it will be necessary to distinguish between various possible treatments. If it is required that the landscape be restored completely to its original appearance and use, then topsoil will be necessary to allow the re-establishment of vegetation similar to that present before mining commenced. Topsoil will also be necessary if a vegetation is required that can be used for agriculture. If, however, it is required that the landscape be only partially restored, or it is only required that the waste be stabilized to prevent erosion and environmental pollution, then topsoil is unnecessarily expensive and detrimental to profitability. In this case tolerant plant populations provide an important alternative to physical and chemical techniques of stabilization already being used. They could be cheaper and will be more satisfactory from the point of view of amenity. An added advantage is that tolerant plant populations, properly handled, could last indefinitely, in contrast to physico-chemical techniques, most of which only last a short time.

References
1.   Antonovics J., Bradshaw A.D. and Turner R.G. Heavy metal tolerance in plants. *Adv. ecol. Res.,* **7**, 1971. 1–85.
2.   Bradshaw A.D. Plants and industrial waste. *Trans. bot. Soc., Edinb.,* **41**, 1970, 71–84.
3.   Dean K.C., Dolezal H. and Havens R. Utilisation and stabilisation of solid mineral wastes. A Progress Report, U.S. Department of the Interior, Bureau of Mines, Pittsburgh, Pa. 1968, 37 pp.
4.   Gadgil R.L. Tolerance of heavy metals and the reclamation of industrial waste. *J. appl. Ecol.,* **6**, 1969, 247–59.
5.   Gregory R.P.G. and Bradshaw A.D. Heavy metal tolerance in populations of *Agrostis tenuis* Sibth. and other grasses. *New Phytol.,* **64**, 1965, 131–43.
6.   Hilton K.J. ed. *The Lower Swansea Valley project* (London: Longmans, Green, 1967), 329 pp.
7.   Jowett D. Adaptation of a lead-tolerant population of *Agrostis tenuis* to low soil fertility. *Nature, Lond.,* **184**, 1959, 43.
8.   Jowett D. Population studies on lead-tolerant *Agrostis tenuis. Evolution,* **18**, 1964, 70–80.
9.   Turner T.G. The subcellular distribution of zinc and copper within the roots of metal-tolerant clones of *Agrostis tenuis* Sibth. *New Phytol.,* **69**, 1970, 725–31.
10.   Wilkins D.A. A technique for the measurement of lead tolerance in plants. *Nature, Lond.,* **180**, 1957, 37–8.

Questions

1   Why is there a need to find both economic and permanent methods for stabilizing toxic mine wastes?
      What methods have been tried and what are the major problems?

2   Grass species such as *Agrostis* and *Festuca* have been used successfully in the stabilization of toxic mine wastes. Why can varieties of these grasses grow in such toxic conditions?
      Why is it much more difficult to use trees instead of grasses?

4.4.7. Pollution of the water*
J.A. Tetlow†

Introduction

There has been an increasing awareness in recent years that strenuous efforts will have to be made, and enormous sums of money spent, in order to protect the nation's water resources from the deleterious effects of widespread pollution and to halt the associated deterioration in the quality of life.

Events such as European Conservation Year 1970 have illustrated the magnitude of the problem. The increasing destruction of amenities and wildlife by population growth and the development of new and potentially dangerous chemicals is now widely recognized. However, for many years River Authorities and related bodies have striven, within the present legal framework, to carry out their statutory duties and protect the environment against pollution.

In this brief paper, the author proposes to outline the present situation of the tidal River Thames, for which the Port of London Authority is the statutory pollution control body, and to include some details of several important historical developments in the conservancy of this famous waterway. The author will then briefly review possible future problems and developments in the field of water pollution control.

Effects of pollution loads

Although . . . the tidal Thames was used as a depository for rubbish from Roman times, the general condition of the river was reasonable until the early nineteenth century. This was because the foul drainage of houses was usually to cess-pits, where the contents were infrequently removed by 'nightmen' and spread on the land. The general introduction of the water-closet in 1810, followed by legislation in 1848 making house drainage obligatory, turned the river into an open sewer. The condition of the river became so offensive that in the 1850s, sessions of Parliament were interrupted and lime-soaked sheets were hung at the windows to reduce the stench. The waterworks were compelled to move their intakes to above the worst affected reaches even though the Chelsea Waterworks Company had installed slow sand-filtration as early as 1829. The deteriorating position was illustrated by the fact that the river was more impure at Battersea Fields in 1849 than it had been in 1832 at London Bridge [see Fig. 1].

There were major cholera outbreaks in the period 1832 to 1886 and in the outbreak of 1849 more than 14 000 people died. Up to this time it was considered that cholera was an airborne disease, but in 1855 Dr John Snow published a paper that illustrated cholera to be waterborne. He described an outbreak in an area bounded by Oxford Street, Regent Street, Piccadilly, and Dean Street where there were more than 500 deaths in 10 days. The source of infection was shown to be a well in Broad Street that had a sewer passing nearby.

In view of this appalling state of affairs the government, in 1855, set up the Metropolitan Board of Works to maintain main sewers and construct new works to prevent sewage entering the Thames in the London area. A scheme of intercepting sewers, pumping stations, and the embankment of

*Reprinted from *Medicine, Science, and the Law*, **12** (2), April 1972, by permission.

†Currently Assistant Director (Water Quality), Anglian Water Authority, Diploma House, Huntingdon PE18 6NZ.

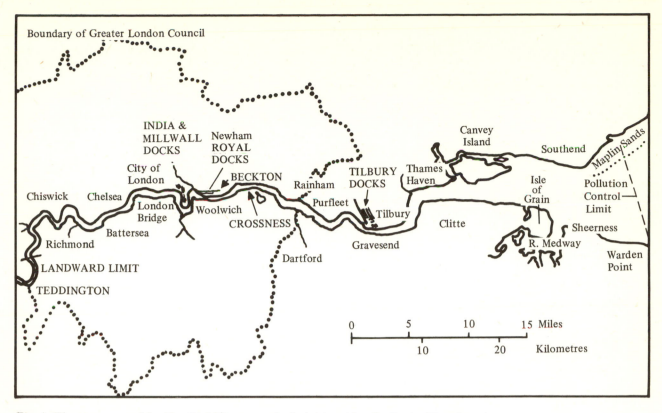

Fig. 1. The area covered by the tidal Thames and administered by the Port of London Authority.

the river in Central London, was drawn up in 1856 by Mr Joseph Bazalgette (later Sir Joseph) and was implemented by an Act of Parliament in 1858; the whole works were completed in 1875. The sewers, which traversed London in a westerly to easterly direction intercepting the main-line sewers on route to the Thames, were to carry the sewage of London to the tidal Thames below the Metropolis. From 1864 drainage from north of the river discharged at the northern outfall at Beckton, and from south of the river discharged at the southern outfall at Crossness, where it was planned to discharge from reservoirs in the early hours of the ebb tide to help carry the sewage out to sea. However, although the condition of the river improved in Central London, there were many complaints regarding offensive conditions from Greenwich to Greenhithe. The Metropolitan Board of Works commenced to install chemical precipitation processes at the northern and southern outfalls to improve the situation, the resultant sludges being transported and dumped at sea. The London County Council, which succeeded the Metropolitan Board of Works, continued to improve treatment facilities as the new biological treatment processes became available around 1915. A new works, at Crossness, capable of giving full biological treatment to the whole of the sewage flow commenced full operation in 1964. The Beckton works, which was able to give full treatment to 60 million gallons per day of the sewage flow before the Second World War, is currently being extended to give full treatment to a dry-weather flow of almost 300 million gallons per day. The present position is that the wastes of approximately 11 million people enter a 20-mile stretch of the tidal Thames and more sewage effluents are added downstream.

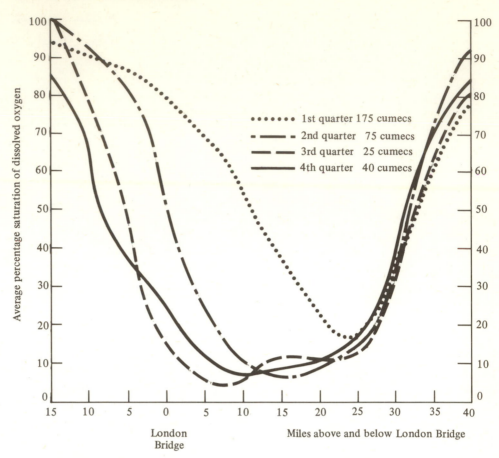

Fig. 2. Typical river oxygen levels related to flow and season.

Effects of early work and technical studies

Improvements in the condition of the river, due to installing the intercepting sewers, did occur as evidenced by the return of fish-life to the river (Cornish, 1902). However, these improvements proved to be only of an interim nature and from about 1920 the condition of the river water continued to deteriorate. Lack of funds for new sewage-treatment plant due to the war-effort, and to bomb damage, resulted in deplorable conditions after the war. About 16 miles of the middle reaches were black, fetid, and liberating hydrogen sulphide. Considerable alarm was expressed regarding public health and amenity aspects and the Port Authority, in order to study the causes of pollution, was instrumental in setting-up the Thames Survey Committee in 1949 under the auspices of the Water Pollution Research Board. There was, after the war, concern also that adverse effects would be caused by the additional power stations being constructed along the river.

The government, therefore, in 1951 appointed a committee, under the chairmanship of Professor Pippard to examine the effect of heated and other discharges on the river.

The Port Authority was fully involved in these two mammoth studies and was able to provide physical flow and sedimentation data from its hydraulic model of the estuary, plus data derived from monitoring major polluting discharges since 1909. Invaluable information was also provided by the London County Council which had carried out weekly chemical surveys of the estuary over the previous 50 years.

144

The investigations proved so complex that the Thames Survey Committee, jointly reporting with the Water Pollution Research Laboratory, did not publish its findings until 1964 (Department of Scientific and Industrial Research, 1964). This report provided the basis for the scientific management of the Thames Estuary and has become a standard work of reference on the behaviour of estuaries receiving pollution loads.

The report of the Pippard Committee, published in 1961 (Ministry of Housing and Local Government, 1961) recognized the prime role of dissolved oxygen in the purification of the pollution loads in the estuary, and recommended that a reserve of dissolved oxygen, sufficient to deal with accidental discharges or storm-overflows, be maintained at all times and in all places. The reports clearly showed the tidal Thames to be virtually an enclosed system and that practically all the pollution loads were exercised in the estuary. The oxygen required for purification processes is 95 per cent derived from the atmosphere and the effect of wind and tides is therefore very important. When there is insufficient dissolved oxygen, aerobic bacteria cannot function and the pollution loads are attacked by anaerobic bacteria that obtain their oxygen first from available nitrates and then from sulphates. The latter produce the foul-smelling gas, hydrogen sulphide, and turn the river water black due to the precipitation of insoluble iron sulphide.

The mechanics of the estuary are such that a slug of polluting matter can travel downstream for 10 miles on the ebb tide, only to return 9½ miles on the flood tide. Thus, depending on the magnitude of the upland flow, polluting matter can remain in the estuary for 6 weeks to 3 months producing a dissolved oxygen 'sag-curve' where the base of the sag can alter its geographical location as the upland flow changes. Fig. 2 illustrates typical dissolved oxygen concentrations relating to flow and season.

The technical investigations evolved the concept, for the Thames estuary, of 'effective oxygen load' (EOL) as owing to the prolonged retention of polluting matter, the long-established Royal Commission standards, dealing with biochemical oxygen demand and the concentration of suspended solids, were inappropriate. The EOL takes into account both the carbonaceous and nitrogenous components of pollution loads and is approximately represented by the equation:

$$E = 1.5 \, (B + 3N)$$

where E is the effective oxygen load, B is the biochemical oxygen demand, and N is the concentration of oxidizable nitrogen. It can therefore be seen that the nitrogenous load has a large effect in the estuary and the removal of oxidizable nitrogen by nitrification processes at the sewage treatment works is of great importance. However, this process requires a high degree of operational control and involves extended aeration with its attendant increased plant and running costs. The investigations showed the pollution loads to the tidal Thames to be comprised as given in Table 1.

In terms of EOL, the present pollution load to the river from sewage treatment works is of the order of 450 tons EOL per day and it is estimated that only 250 tons of oxygen per day are absorbed by the river from the atmosphere in quiescent conditions; the resultant deficit is responsible for the dissolved oxygen 'sag-curve'. As mentioned earlier, high winds can impart large amounts of oxygen into the river water to assist in the purification process. It is estimated that a Force 5 wind will impart about 450 tons of oxygen per day into the river, thus balancing the pollution load derived from sewage treatment works.

Table 1. *Pollution loads in the tidal Thames*

Source of pollution load	Percentage of total load
Sewage treatment works*	79
Direct industrial discharges	12
Upper Thames	4
Tributaries	3
Storm-overflows	2

*Includes trade effluents discharged to sewer.

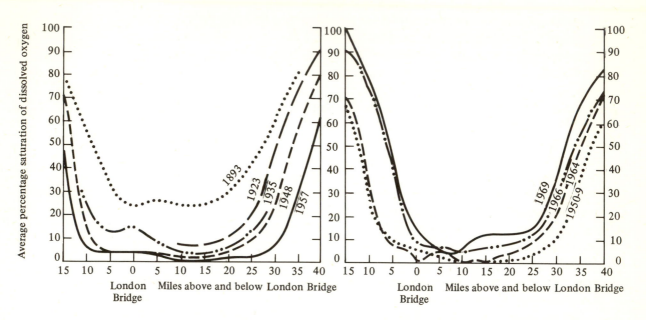

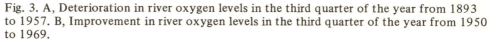

Fig. 3. A, Deterioration in river oxygen levels in the third quarter of the year from 1893 to 1957. B, Improvement in river oxygen levels in the third quarter of the year from 1950 to 1969.

The most critical period of the year is around the third quarter when dissolved oxygen concentrations are at their lowest due to low upland flows and additional loads are being imposed on the river due to colder weather causing increased generation of electricity. Fig. 3 illustrates how the condition of the river deteriorated from 1893 to 1957, and how, due to a positive pollution control policy, the condition has continued to improve since the 1950s.

The Port Authority has adopted a system of consultation, co-ordination, and control in its efforts to reduce the pollution loads to the river. Within this framework the requirements of health, commercial, and amenity aspects are co-ordinated as no one topic should be considered in isolation at the expense of the other two. It is also essential to assure all dischargers that they are being treated in an equitable manner when fixing the standards required for their effluent discharges. Wherever possible, discharges are diverted to local authority sewers, so that the trade effluent is treated at a sewage treatment works with facilities and technical expertise to optimize the purification processes.

By co-ordinating the efforts of local authorities and industrialists, pollution loads were reduced by 17 per cent in the period 1953 to 1962 and by a further 23 per cent by 1969. The largest single contributor to these improvements is the Greater London Council (and its predecessor the

London County Council) which, together with smaller sewage treatment authorities, will have spent approximately £45 million between 1955 and 1973; while industry, to ensure equitable treatment, will have spent between £3 million and £4 million. A result of this expenditure is the collection of sewage flows at large, well-equipped sewage treatment works so that now, in the London area, about 10 large sewage treatment works have replaced about 190 smaller works that existed in 1935.

A tangible way of illustrating the improvements in the condition of the river water is the increase in fish species present in the tidal river. A survey carried out in 1957—8 showed that, with the exception of eels, no fish were found below Richmond downstream to Gravesend, whereas a similar survey since 1967 has shown the presence of some 57 species of fish over the same 40 miles of river (Wheeler, 1969; *Port of London*, 1971). Certain of these species, such as the red mullet, are rare even in the North Sea, and examination of sea-trout taken at Charlton and Teddington earlier this year showed that these fish had travelled through the most polluted zone of the river in their journey from the sea. Associated with the improvements in the condition of the water and increasing marine life are the return of many species of birdlife. A survey in 1969—70 in the Thamesmead area, by the London Natural History Society, counted 1600 shelduck and 330 pintail duck in this one bay alone (Wheeler, 1969). Other species, previously unknown in the river, such as teal, tufted duck, and pochard, were counted in large numbers.

There are, however, some adverse effects due to the improved state of the river such as hydroid growths blocking circulating water intakes to factories and power stations in the lower estuary. The boring marine organism *Teredo navalis*, that can wreak havoc in timber piling and jetty structures, is moving further upstream, having previously been driven out of the estuary by the polluted state of the water.

Future plans

The Port Authority operates a 10-year rolling plan for pollution control activities and has set itself a target of obtaining a dissolved oxygen concentration of 10 per cent saturation at all times and in all places by 1980. Even under the most adverse conditions of the third quarter, a minimum reserve of 10 per cent saturation of dissolved oxygen is planned, and thus at other times of the year the minimum dissolved oxygen concentration will be substantially in excess of this figure. The additional costs in treatment plant and running costs to attain this figure could well amount to a further £40 million.

The tidal Thames has a greater capacity to accept polluting loads as one progresses seawards and therefore different standards are required from discharges into the river at different points. Local authorities may therefore incur very different sewage treatment costs although situated only 20 miles apart. Because of its position at the head of the tidal Thames, the Greater London Council has to bear the greatest burden of sewage treatment costs per head of population.

It is accepted that the heat discharged to the tidal Thames by power station circulating water discharges increases the rate of bacterial activity in the river causing a more rapid depletion of the dissolved oxygen concentration and also it reduces the total quantity of dissolved oxygen that can be dissolved in the river water. To offset such 'thermal pollution' loads, the Port Authority may require that compensatory aeration is provided. Ideally, such compensation schemes should be provided where dissolved oxygen concentrations are at a minimum to obtain maximum returns for money invested in such schemes. However, as the condition of the river improves, the benefits derived from such schemes diminish and therefore their general application is limited. It is obvious that maximum benefit would derive if such aeration was provided at large sewage works in the form of extended aeration to further improve the quality of effluents discharged to the river. It is perhaps, not too Utopian to foresee a day when industrialists may subscribe towards a central pollution control works, that receives free off-peak electricity to provide extended biological treatment. This free electricity would be the power stations' 'compensatory treatment' for their thermal load and would also allow them a more efficient and economic 'base-load' operation.

Additional pollution control problems

As a Harbour Authority, the Port Authority administers the Oil in Navigable Waters Act 1955 and operates a scheme for minimizing the effects of, and removing oil that may be spilt in its area. Facilities are arranged for the reception of sewage ashore from ships in the docks; also houseboat colonies and floating restaurants are required to discharge their sewage to the local authority sewer.

Relatively new materials such as nylon ropes and polythene materials cause a specific problem as they float just below the surface of the water and are caught by vessels' propellers. Due to frictional heat they melt and effectively jam propellers, necessitating the services of a diver to free the offending material.

The Authority has expressed concern regarding the proposed Thames Barrier and has pointed out the dangers of tidal control in relation to water quality. It is essential, in order that sufficient oxygen be drawn from the atmosphere into the water, that there is a certain degree of turbulence. Not only would this be lacking in an impounded area but also the movement of material in and out of that area would be markedly slower. In fact, although the go-ahead has been given to a tidal surge barrier, it will not be operated as a half-tide structure, either continuously or occasionally, until it is established that any problems associated with such control can be overcome.

Since 1887 the sewage sludge from the northern and southern outfall works has been deposited in the outer estuary in the Black and Barrow Deeps. Careful checks are carried out to ensure that there are no serious adverse effects on the marine environment and a recent report (Shelton, 1971) indicated that the sludge was providing a source of food and that fish-life was flourishing. However, there are increasing pressures to dump toxic materials in the seas and the concentration of materials such as poly-chlorinated biphenyls can build up in food chains to become a serious danger to higher forms of life, including man. The Port Authority therefore liaises with other interested organizations and government agencies in carefully monitoring the disposal of any potentially hazardous materials.

Eutrophication is becoming a problem in both inland and coastal waters and is the enrichment of water by plant nutrients. This enrichment has become so great in many waters that excessive growths of algal 'blooms' develop each year and, when such waters are used for supply, problems of taste and odour sometimes arise (Owens, 1970). Loss of efficiency of filters used for water treatment, through accelerated clogging, also occurs. These blooms destroy the amenity aspects of reservoirs and can so take up dissolved oxygen that fish become asphyxiated. Certain types of blue-green algae can

148

also produce substances that are toxic to fish and other animals. Recent studies indicate that about 80 per cent of the phosphorus but less than 20 per cent of the flow of nitrogen, silicon, potassium, chloride, and sulphate was derived from sewage effluent, the remainder being introduced in the run-off from agricultural land. Thus, phosphorus could be substantially reduced by chemical removal at the sewage treatment works, but it would not be economic to remove nitrogen at the same time as the bulk of the nitrogen is derived from agricultural run-off. However, if nitrogen presented a problem to potable water sources it would be feasible to remove it at the water-abstraction point (Bayley, 1970).

New developments

Probably the most acute problem in the present conventional sewage treatment processes is that of sludge disposal. The long-established technique using sludge drying beds is now obsolete, because of the shortage of the large land area required, particularly in built-up areas. Many new mechanical techniques have been developed for sludge dewatering and pressing to replace the sludge drying bed. In many cases a saleable commodity is produced that can be used as a cheap fertilizer. However, the increasing presence of toxic chemicals in sewage sludges renders the end-product unfit for agricultural use. In such cases, incineration techniques will be of increasing importance in the future and the multiple-hearth and fluidized bed furnaces are now feasible propositions. A novel new technique is the vacuum system developed in Sweden (Ministry of Housing and Local Government, 1970) which uses air instead of water for the transport of sewage and, therefore, allows the separation of 'black-water' (faecal matter and urine – lavatory waste) from 'grey-water' (all other household liquid wastes).

The concentrated 'black-water' is treated chemically, which removes substantial quantities of the nutrients nitrogen and phosphate, as the 'black-water' contains 91 per cent and 42 per cent respectively of these constituents of the total found in domestic sewage. The 'grey-water' is treated biologically as it has been shown that it is easier to treat this biologically than either 'black-water' or a mixture of the two components. The saving in domestic water consumption with this system is claimed to be about 30 per cent and there would be a corresponding welcome reduction in the volume of domestic sewage discharged to the sewers; thus reducing expenditure on new sewerage systems.

Future organization of water resources

Changes are urgently required in the management of the country's water resources due to an acute water shortage in certain parts of the country. Areas of high population density are, unfortunately, located in areas of low rainfall. The Water Resources Board is implementing several schemes, and considering others, to alleviate the problem. The building of new reservoirs is often opposed by many conservation lobbies, although river regulation schemes have many points of merit. The transfer of water between hydrometric areas, as in the Ely/Ouse Scheme, will increase and barrages may be built in Morecambe Bay and The Wash to retain fresh water that would otherwise be lost to sea at times of flood. There will be increasing re-use of sewage effluents by industry and new techniques, although at present costly, such as reverse osmosis, can produce potable quality water from a good sewage effluent.

Three committees have recently published reports that will have far-reaching effects on the future management of water resources in England and Wales. In March 1970, the Report of the Working Party on Sewage Disposal, *Taken for Granted* was published (Ministry of Housing and Local Government, 1970). The Working Party was appointed in February 1969, 'to consider and report on the public health, amenity, and economic aspects of the various methods of sewage disposal'. Recommendations included the improvement in education and training facilities for personnel engaged in the water pollution control field, and that managers of sewage treatments works should be required to hold the appropriate professional qualifications.

A Royal Commission on Environmental Pollution under the Chairmanship of Sir Eric Ashby, stated in its first report in February 1971, that its first priority was 'to enquire into and report on the problems of pollution of tidal waters, estuaries, and the seas around our coast'. The Commission also made recommendations concerning the qualifications and training of those who control water pollution, and on the future administration of rivers and sewage treatment.

The most recent report is that of the Central Advisory Water Committee (1971) that was reappointed in September 1969, 'To consider in the light of the Report of the Royal Commission on Local Government in England and of technological and other developments how the functions relating to water conservation, management of water resources, water supply, sewerage, sewage disposal and the prevention of pollution now exercised by river authorities, public water undertakings, and sewerage and sewage disposal authorities can best be organized; and to make recommendations'. This was obviously a very difficult task, and because of diverging opinions the Committee could not arrive at unanimous recommendations. Alternatives for the organization of the four basic functions of water-supply, sewage disposal, river management, and planning and co-ordination were presented either by multi-purpose authorities, or some system based on single-purpose authorities. The government is now considering the recommendations of the Committee and a White Paper is expected to be published soon.* Whichever of the schemes is chosen there will be drastic alterations in the system of management of water in England and Wales, and it is hoped that the new organization will be adequate to cope with the severe problems in this field for many years to come.

Conclusion

Rapid advances have been made in the sphere of pollution control in recent years, but technological innovations such as the development of new materials and chemicals, plus a predicted population increase in Britain from about 56 million in 1971 to about 69 million in the year 2000, leave no room for complacency. The expertise exists to control pollution provided sufficient funds are made available; as these are, in one way or another, provided by the public, it is essential that the public is adequately informed of the situation. It is estimated that pollution control measures urgently required in the United States of America will add 5 to 10 per cent on to the Cost of Living Index. To quote an example nearer home, it could cost many millions of pounds in additional sewage and trade effluent schemes to ensure that salmon can once again pass up the River Thames to London Bridge. Thus, in the foreseeable future it is obviously uneconomic and a compromise will be

*Government proposals now published in Department the Environment Circular No. 92/71.

150

reached that balances the commercial and amenity requirements of this famous waterway. However, the example illustrates that the choice is a subjective one, and that the pollution control authority must always plan and operate in advance of informed public opinion.

Thanks are due to Mr J.H. Potter, Chief Engineer, River and Principal Pollution Control Officer, Port of London Authority, for permission to present this paper.

References
Bayley, R.W. (1970). *Water Treatment and Examination,* **19**, 294.
Cargill, C. (1969). *The River Thames—Historical Survey of the Rights over the River and their Conservancy*. London: The Thames Angling Preservation Society.
Central Advisory Water Committee (1971). *The Future Management of Water in England and Wales*. London: H.M.S.O.
Cornish, C.J. (1902). *The Naturalist on the Thames*. London: Seeley.
Department of Scientific and Industrial Research (1964). *Effects of Polluting Discharges on the Thames Estuary*. Water Pollution Research Laboratory Technical Paper No. 11. London: H.M.S.O.
Ministry of Housing and Local Government (1961). *Pollution of the Tidal Thames*. London: H.M.S.O.
− − (1970). *Taken for Granted*. Report of Working Party on Sewage Disposal. London: HMSO
Owens, M. (1970). *Water Treatment and Examination*, **19**, 239.
Port of London, June (1971), The Living Thames.
Royal Commission on Environmental Pollution (1971), *First Report, Cmnd.* 4585. London: HMSO.
Shelton, R.G.J. (1971), *Marine Pollution Bulletin*, **2**, 24.
Wheeler, A. (1969), *Biol. Conserv.* **2**, 25.

Questions

1 Why is the percentage saturation of dissolved oxygen a useful indicator of river pollution?
2 What factors (both physical and biological) will determine the percentage saturation of dissolved oxygen at any given time?
3 What methods and measures have been adopted by the Port of London Authority to improve the Thames water since 1957?

4.5. Questions relating to the objectives of the book

When you have worked through the book, test yourself with these questions to see whether you have mastered the objectives set.

1   Considering trophic levels n, $n-1$, $n-2$, and $n-3$ in a food chain, which of the following is a measure of gross ecological efficiency?

(a) $\dfrac{\text{The amount of } n-1 \text{ consumed by } n}{\text{The amount of } n-2 \text{ consumed by } n-1} \times 100$

(b) $\sqrt{\dfrac{\text{The amount of } n-2 \text{ consumed by } n-1}{\text{The amount of } n-3 \text{ consumed by } n-2}}$

(c) $\dfrac{\text{The change in standing crop of } n}{\text{The change in standing crop of } n-1} \times 100$

(d) $\dfrac{\text{The rate at which the standing crop of } n-1 \text{ is converted to } n}{\text{The rate at which the standing crop of } n-2 \text{ is converted to } n-1} \times K_f$

(e) $\dfrac{\text{The amount of } n \text{ standing crop lost as heat}}{\text{The amount of } n-1 \text{ standing crop lost as heat}} \times K_f$

(f) $\sqrt{\dfrac{\text{The rate of production of } n \text{ minus the rate of production of } n-1}{K_f}}$

[*Objective* 1, section 4.2]

2   Which of the following are features of a mature ecosystem?
(a) High species diversity
(b) High production efficiency
(c) High biomass stability
(d) Slow exchange of nutrients between organisms and their environment
(e) High level of stratification
(f) Broad niche specificity
(g) Detritus important in nutrient regeneration
(h) $\dfrac{P}{R} \simeq 1$
(i) Growth forms characterized by feedback controls

[*Objective* 1, section 4.3]

3   Which of the following interactions will occur at *all* stages of a *developing* ecosystem?
(a) Competition
(b) Climax
(c) Colonization
(d) Invasion
(e) Succession

[*Objective* 1, section 4.3]

4   Consider the following examples of feeding.
   (*a*)   A worm living in the shell of a hermit crab, draws into its mouth
           particles of food broken off from the crab's meal.
   (*b*)   A worm living in the blood vessels of a man's leg, feeds on the
           blood cells causing sickness and debility to the man, but not death.
   (*c*)   A worm feeds on the fallen leaves of a beech tree.
   (*d*)   Bacteria in the gut of a worm secrete chemicals which digest the
           cellulose in the diet of worm (*c*), thus providing it with sugars. The
           bacteria in turn feed on the products of the worm's digestive
           system. No harm accrues to either organisms.
   (*e*)   A thrush feeds on a worm.
   Pair each of the above feeding relationships with the appropriate
   descriptive term chosen from the list below:
   (i)   Symbiosis
   (ii)   Saprobism
   (iii) Predation
   (iv)   Commensalism
   (v)   Parasitism

                                        [*Objective* 2, section 4.2]

5   Ecological pyramids show the relationship between:
   (*a*)   biomass and trophic level
   (*b*)   number and size of organisms
   (*c*)   number of organisms and trophic level
   (*d*)   number of organisms and number of species in the ecosystem
   (*e*)   taxonomic groupings and the distribution of species within one
           ecosystem.

                                        [*Objectives* 1 and 2, section 4.2]

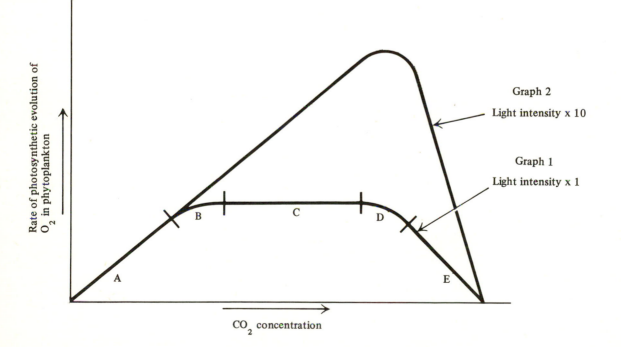

6   Which of the following are true observations about the section of the
      graph indicated? (*Note.* Often more than one are correct)

                                                                      153

Section A   (i)   Light is a limiting factor.
                (ii) CO_2 is a limiting factor.
                (iii)   The pH is too low for optimum growth.
                (iv)   There are no limiting factors.
                (v)   The rate of photosynthesis is proportional to the CO_2 concentration.

Section B   (i)   Light is a limiting factor.
                (ii) CO_2 is a limiting factor.
                (iii)   Water is a limiting factor.
                (iv)   The pH is becoming too low for optimum growth.
                (v)   The rate of photosynthesis is maximal.

Section C   (i)   Light is a limiting factor.
                (ii) CO_2 is a limiting factor.
                (iii)   The pH is too low for optimum growth.
                (iv)   The phytoplankton are suffering from oxygen-poisoning.
                (v)   There are no limiting factors.

Section D   (i)   Light is a limiting factor.
                (ii) CO_2 is a limiting factor.
                (iii)   The pH is becoming too low for optimum growth.
                (iv)   Nitrate and phosphate are becoming limiting as they are being used up.
                (v)   The temperature is moving from the optimum for enzymic activity.

Section E   (i)   Light is a limiting factor.
                (ii) CO_2 is a limiting factor.
                (iii)   The pH is too low for optimum growth.
                (iv)   The temperature change is beginning to denature the enzymes.
                (v)   The phytoplankton are dying off.

[*Objectives* 1, 3 and 4, section 4.1]

7   Which of the following are ecosystems?
     (i)   a self-contained spacecraft
     (ii)   the zooplankton and their decomposers in a lake
     (iii)   duckweed floating·on the surface of pond water in a beaker containing decomposers
     (iv)   all the *Daphnia* (water-fleas) in a pond and their decomposers
     (v)   a ditch with all its living and non-living components
     (vi)   a culture of phytoplankton in a sterile culture medium

[*Objective* 1, section 4.2]

8   When using insecticides like DDT, which of the following are true and which false?
     (i)   DDT is a persistent organo-chloride which can remain undegraded in the ground for 10–15 years.
     (ii)   The concentration of DDT decreases as it passes through a food chain.
     (iii)   Spraying harmful insects like mosquitoes with soluble DDT is quite justified provided that the spray concentrations are not directly lethal to fish and other wildlife.
     (iv)   It would be less harmful to use DDT to eradicate mosquitoes in a

stagnant pond than using it to eradicate pests in a fruit orchard.
[*Objective* 1, section 4.4]

9 Match correctly each of the following lettered terms or statements with one of those numbered.

A. Frogbit and duckweed
B. Eutrophication
C. Starwort and water forget-me-not
D. Ecological niche
E. Autotrophic
F. Copper and lead mines

i. Trophic position in the community
ii. Selectively tolerant plants
iii. Free-floating plants
iv. All photosynthetic organisms
v. Emergent plants
vi. Excessive quantities of phosphates and nitrates

[*Objectives* 3, section 4.4
1, section 4.3
2, section 4.2
2, section 4.1]

10 The following shows the energy flow through a small portion of a grassland ecosystem. The figures given are in kilojoules/m² per year.
N.P. = Net production
G.P. = Gross production

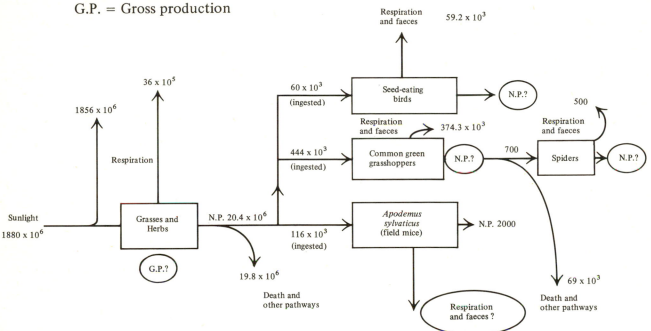

Questions

(i) What is the photosynthetic efficiency of the system?
(ii) What is the gross production of grasses and herbs?
(iii) What is the net production of the seed-eating birds, spiders and common green grasshoppers respectively?
(iv) How much energy is lost via respiration and faeces by field mice?
(v) If the total respiratory energy lost as heat is 14×10^6 kJ is the ecosystem close to maturity?
(vi) Which of the labelled boxes are producers, 1° consumers, 2° consumers?
(vii) Which of the labelled boxes are heterotrophic organisms?
(viii) What are the 'other pathways' likely to be? (Name three.)

155

(ix)   Which of the labelled boxes are communities?
(x)   Which of the labelled boxes are populations?

[*Objectives* 2, section 4.1
1 and 2, section 4.2
1, section 4.3]

11   The following graph shows the present relationship between four parameters; environmental impact (E), world population (P), average living
standards (L; i.e. per capita consumption of resources and space) and
world resources (R; i.e. energy, materials, space). The axes given are the
relative magnitudes (log scale) of each parameter against time.

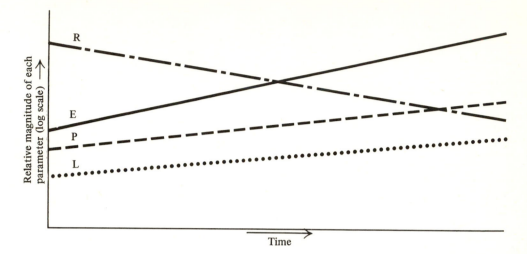

Predict the effect on the other parameters
(i)   if the world population remained stable at its present level, but
the average living standard continued to increase;
(ii)   if the world population continued to increase but the present
living standards remained unaltered and we recycled half of our
material resources;
(iii)   if the impact on the environment remained stable at its present
level, but the average living standards increased;
(iv)   if the environmental impact and the world population remained
stable at their present levels.

[*Objectives* 1 and 2, section 4.4]

12   Which of the following have been important in alleviating pollution of
the river Thames?
(*Choose more than one.*)
(i)   Maintaining reserves of dissolves oxygen to deal with accidental
discharges and storm overflows.
(ii)   Installation of intercepting sewers.
(iii)   Weekly chemical surveys.
(iv)   Removal of oxidizable nitrogen by nitrification processes.
(v)   Co-ordinating the efforts of local authorities and industrialists.
(vi)   New sewage treatment plant.
(vii)   Extensive funding.

[*Objective* 3, section 4.4]

Answers to self-assessment questions

1   *a.*
2   *a, c, d, e, g, h, i.*
3   *a, c, d, e.*
4   *a*(iv); *b*(v); *c*(ii); *d*(i); *e*(iii).
5   *a.*
6   Section A (ii), (v).
    Section B (i), (ii).
    Section C (i).
    Section D (i), (iii).
    Section E (i), (iii).
7   (i), (iii), (v).
8   (i) true; (ii) false; (iii) false; (iv) false.
9   A(iii); B(vi); C(v); D(i); E(iv); F(ii).
10 (i) 1.3%; (ii) 24 X 10⁶ kJ/m² per yr.; (iii) 800, 200 and 69 700 kJ respectively;
    (iv) 114 X 10³ kJ; (v) No; $\overline{\frac{P}{R}}$ = 1.7; (vi) grasses and herbs (producers), seed-eating
    birds, common green grasshoppers and field mice (1° consumers), spiders (2°
    consumers); (vii) all except grasses and herbs; (viii) other primary or secondary
    consumers, decomposers, emigration; (ix) grasses and herbs, seed-eating birds and
    spiders; (x) the common green grasshoppers and *Apodemus sylvaticus.*
11 (i) There will still be some increase in environmental impact and demand on
         resources.
    (ii) Some reduction in depletion of resources but impact on environment would
         still increase.
    (iii) The population level must go down.
    (iv) The average standard of living must go down.
12 All these factors have been important.

157

4.6. Recommended reading

A Blueprint for Survival.   Reprinted from *The Ecologist.* Penguin, London (1972).

Dartnell, R.M. (1973)   *Ecology and Man.* Wm C. Brown, Dubuque, Iowa.

Erhlich, P.R. & A.H. (1972)   *Population, Resources, Environment: Issues in Human Ecology*, 3rd edition. W.H. Freeman, San Francisco.

Lerner, I.M. (1968)   *Heredity, Evolution and Society*. W.H. Freeman, San Francisco.

Odum, E.P. (1963)   *Ecology*. Holt, Reinhard & Winston, New York.

Odum, E.P. (1971)   *Fundamentals of Ecology*, 3rd edition. W.B. Saunders, Philadelphia.

Phillipson, J. (1971)   *Ecological Energetics.* Edward Arnold, London.

Index

Masking card

Objectives of the book

Whilst working through the text please bear in mind the following, which you should be able to carry out on completion of the relevant sections.

4.1

1. Formulate and explain the concept of limiting factors.
2. Formulate and explain the concept of production.
3. Demonstrate some skill in
   (a) formulating hypotheses;
   (b) designing experiments;
   (c) interpreting graphical data.
4. List and describe the basic requirements which plants have of their environment.

4.2

1. Explain, giving examples, the following:
   (i) the concept of an ecosystem;
   (ii) the trophic interactions between organisms within an ecosystem.
2. Define the following terms:
   autotrophic, heterotrophic, parasite, saprobe, commensal, predation, population, community, photosynthetic efficiency, ecological efficiency, gross production, net production, standing crop, biomass.
3. Demonstrate some skill in the interpretation of reported, graphical and numerical data.

4.3

1. Explain, giving examples, the following:
   (i) the processes of invasion, colonization, competition and succession;
   (ii) the distinctive characteristics of mature as compared with immature ecosystems;
   (iii) the concept of an ecological niche.
2. Demonstrate some skill in the interpretation of reported, graphical and numerical data.

4.4

1. Describe giving examples, the consequences of mismanagement of ecosystems by man.
2. Explain the implications of an exponentially increasing human population on the demand for food, space and resources.
3. Describe two methods by which scientists are trying to improve the quality of some parts of our polluted environment.